Let There Be Life!

By
Noel Brooks

LET THERE BE LIFE!

Copyright 1975 by
ADVOCATE PRESS
Franklin Springs, Georgia

Printed in U.S.A.

Let There Be Life!

In this latest book by Rev. Noel Brooks, he brings his keen analytical insight to bear on a highly controversial subject—divine healing.

In many of the areas examined, he could justifiably have written in strictly negative terms. But he didn't.

Instead, he has depicted the healing ministry of the compassionate Jesus.

In his graphic and scholarly manner he helps us "see" the Stranger of Galilee ministering to the needs of the masses of men and women He touched during His earthly ministry. But more importantly, we behold the rays of Messiah's glory shining out toward the spiritual and physical infirmities of lost men in a darkened world, saying: "Let there be Light—Let there be Life."

CONTENTS

Chapter I.—Defending the Truth 7
Chapter II.—Clarifying the Truth 40
Chapter III.—Interpreting the Truth 64
Chapter IV.—Appropriating and Communicating the Truth 92
Bibliography 116

Chapter I

DEFENDING THE TRUTH

The doctrine of divine healing as taught in the Pentecostal Holiness Church is laid down in Article 12 of the Church's "Articles of Faith" (*Manual of the Pentecostal Holiness Church,* 1973, pp. 17, 18). It reads, "We believe in divine healing as in the atonement," and it adds, in brackets, five Bible references (Isaiah 53:4, 5; Matthew 8:16, 17; Mark 16:14-18; James 5:14-16; Exodus 15:26). This article has the distinction of being the briefest of all our Articles of Faith. It comprises only nine words. The number of Bible passages added in support of the doctrine is surprising in view of its brevity, when compared with other doctrines. No scripture references are given for the first five of our articles (articles which came to us mainly from the Reformation in Germany and England, and are the ground work of all conservative evangelicalism). The doctrine of heaven and hell has three references; the doctrine of atonement three; the doctrine of justification one; the doctrine of sanctification in two articles, has three; the doctrine of the pentecostal baptism has six, five of which are from the Book of Acts; the doctrine of the second coming of Christ, five.

Of course, these Bible references are only samples from a much larger store of biblical material. The point I wish to make is that it is surprising that the most briefly stated of all our doctrines should have such a proportionately large number of Bible references added in its support, being exceeded by only one, the doctrine of the pentecostal baptism, which has six references, and equalled by one other, the doctrine of the second coming of Christ.

The doctrine of divine healing, as defined in our Articles of Faith, begs several questions. For example, *what exactly is meant by "Divine Healing"?* This is a term that is variously understood, and our article does nothing to help us in selecting one view in preference to another, though the expositions of Bishops J. H. King and J. A. Synan later in the *Manual* or *Discipline* (pp. 23 and 35), help us to some degree. Then what is meant by the expression *"as in the atonement"?* In my opinion, one of the most necessary and urgent tasks, not only for the Pentecostal Holiness Church, but for all earnest believers in the doctrine of divine healing, is to come to grips with this problem, and formulate a truly Biblical statement of what is meant by "healing in the atonement."

Our article says nothing else relative to divine healing, yet we know that other important matters are involved. What about the question of *"miracles"?* What about *"faith"?* What about the *"causes of ill health"* and the *"conditions of healing"?* What about the *"will of God"?* What about the *"gifts of healing"* and the *"ministry of healing"?* What about the *"will of God"?* What about never receive healing, in spite of real faith and holiness, and sometimes in spite of the fact that they themselves may have been used by God in the ministry of healing? Our Article of Faith is silent about these matters.

In some ways, therefore, our doctrinal statement leaves much to be desired. We might have been spared much confusion and conflict if we had been given clear guidance about these matters in our Articles of Faith.

Nevertheless, there have been definite advantages from this deficiency. The way has been left open for thought and reflection, even for experiment. Quite possibly our founding fathers had not themselves thought through all the problems and difficulties involved. They themselves were not in a position to lay down the doctrine in very precise terms. It is perhaps a mark of their wisdom, even, maybe, a mark of the Providence of God, that they left us the doctrine as we have it. Nevertheless, they bequeathed us a responsibility. In the light of reflection and experience, some of which has been with heartache and agony, we are responsible to define what we mean by *"Divine healing in the atonement"* as well as to speak plainly and authoritatively upon some of the questions we have mentioned.

That the Living God is concerned about man's physical and mental welfare is unquestionable if we accept the Bible as the revelation of God's mind and will. The ancient Gnostic idea that "matter" is essentially evil and totally outside the divine concern is an impossible concept in the light of the incarnation of the Son of God (John 1:14; 1 John 4:1-6; 2 John 7), and the "redemption of the body" at His Second Advent (Romans 8:23; 1 Corinthians 15:51-54; Philippians 3:20, 21).

God's willingness to heal the human body is plainly revealed in Scripture.

It is plainly stated in specific promises. "If thou wilt diligently hearken to the voice of the Lord thy God, and wilt do that which is right in his sight, and wilt give ear to his commandments, and keep all his statutes, I will put none of these diseases upon thee, which I have brought upon the Egyptians: for I am the Lord that healeth thee" (Exodus 15:26).

"And ye shall serve the Lord your God, and he

shall bless thy bread, and thy water; and I will take sickness away from the midst of thee" (Exodus 23:25).

"Bless the Lord, O my soul, and forget not all his benefits . . . who healeth all thy diseases" (Psalm 103:2, 3).

"Is any sick among you? Let him call for the elders of the church, and let them pray over him, anointing him with oil in the name of the Lord; and the prayer of faith shall save the sick, and the Lord shall raise him up; and if he have committed sins, they shall be forgiven him. Confess your faults one to another, and pray one for another, that ye may be healed" (James 5:14-16).

It is abundantly illustrated by many examples in the Old Testament and in the New Testament. There are a number of clear cases of healing recorded in the Old Testament (Genesis 20:17, 18; 21:1, 2; Numbers 12:9-16; 21:6-9; 1 Kings 13:6; 17:17-23; 2 Kings 4:32-37; 5:10-14; 13:21; 20:4-7; Job 42:10).

The ministry of Christ was very largely (though certainly not exclusively) a campaign against sickness and disease. There are about thirty specific healing miracles recorded in the Gospels. In addition to these individual cases, some passages describe mass healings by Christ (Matthew 4:23, 24; 8:16, 17; 9:35; 14:35, 36; 15:30, 31; Mark 6:53-56; Luke 7:21, 22). It is also said that these were only samples out of a much larger number (John 20:30; 21:25). Moreover, Christ healed through the instrumentality of His twelve apostles during the Gospel period (Luke 9:1-6), and even through seventy other disciples (Luke 10:1-12, 17-20).

This healing ministry of Christ through His followers was continued and greatly amplified after His resurrection and ascension. There are many examples of it in the book of the Acts of the Apostles (3:2; 5:12-16; 6:8; 8:6, 7; 9:17; 9:34; 9:40; 14:8-10; 16:16-18; 19:11; 20:10; 28:8).

Moreover, the fact that the "gifts of healing" are mentioned as part of the church's ministry seems clearly to imply that such an extension of Christ's ministry of healing through the presence and power of the Holy Spirit, was purposed to be an abiding reality in the "church which is his body" (1 Corinthians 12:8).

That there are problems and difficulties in regard to divine healing is not denied. I have tried to face some of these perplexities in my book, *Sickness, Health and God*. Some of them will be discussed in the present lectures. It is not possible, in four brief lectures, to cover all the territory involved. The time element forces selectivity upon us.

In the first lecture I shall try to face up to some of the more serious criticisms which have been brought against the concept of miraculous divine healing in general, both those made by out-and-out rationalists, and those made by Bible-believing evangelical Christians. In the second lecture, I myself shall take a critical attitude (I hope, constructively) to some of the confusions and distortions of Biblical divine healing, which are a source of embarrassment to many earnest Christians.

In the third lecture, I shall try to come to grips with what I believe to be a feature of major importance—a true understanding of what is meant by the term, "healing in the atonement." And in the final lecture, I shall seek to discuss the enormously important and practical question: How can we receive divine healing, and how can we become instruments in the hands of the Risen Christ, for the communication of healing to others?

We are utterly dependent upon the Lord's help as we proceed in our quest, and we feel our great need of the illumination and empowering of His Holy Spirit.

First, Rationalistic Criticism

Those who have taken time to attend this particular series of lectures are not likely to be rationalistic critics. Moreover, people who are rationalistic critics are not likely to incline their ears to what we are saying here. It would be foolish to give more than passing attention to the problem, yet it would be dishonest to say nothing at all.

Rationalistic criticism may be said to be of two broad categories: atheistic, and theistic. Atheistic rationalism denies the existence or operation of any power outside natural law. Nature is a closed system, a self-originating and self-operating machine. On this hypothesis there can be no miracle, no revelation, no answers to prayer, no redemption, nor, indeed, any God who effects these things. Theistic rationalism, on the other hand, postulates a divine Being, but confines Him to the world of natural law. He created the laws of nature. They are the manifestations of His power and wisdom. However, this is the only way He does operate or can operate. He operates naturally, but not supernaturally. The logical conclusion of this philosophy would appear to be Pantheism.

Both these types of rationalistic criticism reject miracles and a supernatural order. Thus they are in conflict with Bible religion, both of Old Testament and New Testament varieties. It goes without saying that there can be no reconciliation between Bible religion and atheistic rationalism.

There is, however, a point of contact between Bible religion and theistic rationalism. The Bible, too, declares that God created the laws of nature and that He continues to uphold them. In fact, without His providential control the laws of nature could not continue in operation. But while declaring that God created natural law and operates within natural law, the Bible cries out loud against the idea that God is wholly contained within natural law, and that His activity is exclusively confined to it.

The God of the Bible is *living, almighty, sovereign;* and does, if His wisdom and love require it, act independently of natural law to fulfil His holy will.

Lax of Poplar, a famous English Methodist, was once honored to be the Mayor of the London Borough of Poplar. During his year of office he was riding with King George V in a stately carriage pulled by a pair of fine horses. The King asked Mr. Lax why, if it was the custom of British Methodism to move ministers to new circuits and churches every three years, he had been minister of the Poplar mission for about twenty years. "Well, Your Majesty," replied Lax, "the Methodist Conference is a wise body. It drives a coach and pair through its own regulations sometimes."

God has set up natural law. He works within natural law. It is a wonderful thing to recognize this. It would save us from fanaticism in the matter of divine healing if we recognized it. We should understand that there is no more inconsistency between faith and the taking of medicine than there is between faith and the taking of food, or of going to sleep. Bishop J. H. King clearly recognizes this in his statement on a natural law of healing in his exposition in the *Manual.* "The law of recovery is written in all creation, and also in our bodies, since they are an essential part of creation. This law operates according to its relation to the infinite law of all creation as upheld and directed by the Creator. Healing is a part of the benefits flowing out of this law of recovery, and it may be termed the healing of natural law" (p. 23).

But the Living God is not a prisoner within natural law. *He can act independently of it.* He can intervene in, and overrule, and transcend, natural law at any moment in order to fulfill His gracious purposes.

Sometimes He does this in answer to the prayers of His people for the healing of their bodies. He did so in Bible days. The Bible encourages us to believe that He is willing to do so today. There have been many people, and there are many people today, who are convinced that God has acted in this fashion for them.

It is my earnest belief that faith in God, in the Biblical meaning of the word, *demands* belief in the reality of His supernatural and miraculous action, independently of natural law. I cannot personally think of God without thinking of such a Supernatural Being. Dr. James Orr says, in his great book *The Christian View of God and the World,* "Many people speak glibly of the denial of the supernatural, who never realize how much of the supernatural they have already admitted in affirming the existence of a personal, wise, holy, and beneficient Author of the Universe. They may deny supernatural actions in the sense of miracles, but they have affirmed a supernatural Being on a grand scale and in a degree which casts supernatural action quite into the shade. If God is a reality, the whole universe rests on a supernatural basis. A supernatural presence pervades it; a supernatural power sustains it; a supernatural will operates in its forces; a supernatural wisdom appoints its end. . . . If the opposition to the supernatural is to be carried out to its logical issue, it must not stop with the denial of miracle, but must extend to the whole theistic conception" (p. 92).

We ought not to be surprised at the "death of God" theology, which is a product of the new liberalism. A theology which denied miracles and the supernatural order, was bound eventually to lead to the denial of a personal God, which is at the heart of the concept "death of God."

Dr. A. B. Bruce, in the final chapter of his work, *The Miraculous Element in the Gospels*, envisages a *Christianity without miracle,* and declares that if this is so, "the days of Christianity are numbered; . . . Christianity without miracle means Christianity ceasing to be a substantive religion, and becoming a mere unnamed element taken up, in so far as good, into some other modern religion bearing a new name, as the soil formed by decomposition of the vegetation of early geological epochs enters into the living plants of the present era" (pp. 375-6). He goes on to say, "When this happens, Christianity, done to death by unworthy faith and by scientific unbelief abhorrent of the supernatural, will repeat the miracle of the resurrection, and run a new career, fraught with glory to Jesus and with manifold blessing to men" (page 388).

In recent years a more specific type of rationalistic criticism has appeared which attacks the miraculous Christian experience of conversion, and all that conversion entails. About a decade ago a book written by an English doctor, Doctor William Sargent, made a challenging impact upon many minds and created somewhat of a sensation in the theological world. It was called *Battle for the Mind,* and bore the sub-title, *A Physiology of Conversion and Brainwashing.* In this book Dr.

Sargent set out to prove that the phenomena associated with religious conversion and political brainwashing are very much alike, and that both can be explained by a purely physiological process such as was demonstrated by the famous Russian scientist, Pavlov, in his experiments on the brains of dogs.

Pavlov discovered that if dogs were subjected to fears, anxieties and tensions through electric-shock treatment or drugs, a state of nervous excitement and collapse could be brought about in which their usual responses to stimuli would be totally inhibited. He then found that he could put totally fresh suggestions into the minds of the dogs which would bring about entirely new "conditioned reflexes" and lead to entirely new patterns of behaviour. During the Second World War Dr. Sargent and others found that the methods which they were successfully using to heal shell-shocked and battle-weary patients, and restore them into normal citizens, were strikingly similar to the methods of Pavlov with his dogs. About this time Sargent became interested in the *Journals of John Wesley,* particularly his narratives of religious revival and conversion. Comparing the techniques of Pavlov and the World War II doctors, he reached the conclusion that what really happened in Wesley's meetings was the same thing which happened to Pavlov's dogs under electric-shock treatment, and to the shell-shocked soldiers when treated with drugs. He then went on to study the brainwashing techniques of the Communists and decided that precisely the same thing occurred.

This thesis, as far as it affects the Christian faith, is shockingly blunt. Christian conversion, he claims, is explainable in these physiological terms. Before conversion can take place, people must be subjected to emotional atmospheres which induce a state of tension and conflict, fear and anxiety. Once begun, the excitement and tension must be built up until inhibition of the rational processes takes place, and finally a collapse, as in Pavlov's dogs. Then new suggestions, ideas, and beliefs can be implanted in the mind, leading to changed patterns of behaviour and a new character or person. What Christians have always believed to be a miraculous work of the Holy Spirit is, according to this theory, wholly explainable in these physiological, mechanistic, behavioristic terms. Dr. Sargent mentions, as methods used by revivalists to achieve this hell-fire and second-coming preaching, rhythmic music, religious dancing, as well as snake-handling. As examples he particularly refers to the early Quakers, Wesley, Finney, the American snake-handlers, and Billy Graham. He also puts the techniques of heathen witchdoctors and the Catholic Inquisition into the same category. Furthermore, he even claims that Peter used these brainwashing methods on the Day of Pentecost to soften-up the crowd preliminary to their conversion. He claims, too, that Saul of Tarsus was converted in the same way. He does not give special attention to faith-healers, but makes the significant statement, "Many kinds of spiritual healers use the same basic technique with differing interpretations added" (*Battle for the Mind*, p. 109).

What are we to say to all this? It is, of course, utterly impossible for a Christian to accept it. Sargent is evidently a "natural man who understandeth not the things of the Spirit of God" (1 Corinthians 2:14). He shows no insight at all into the spiritual order of the kingdom of God. His is a typically rationalistic attitude. His book has been examined and appraised critically by Dr. Martin Lloyd-Jones, in a booklet called, *Conversion: Psy-*

chological and Spiritual. I should like to give one or two quotations from this little book. Dr. Lloyd-Jones declares, "The answer to the suggestion that pervades this book is that the explanation of the events and experiences such as Pentecost, the conversion of the Apostle Paul, John Wesley and others, is not psychological but, always and essentially, theological." He further says, "What Dr. Sargent unfortunately does not realize is that the power of Peter's preaching on the Day of Pentecost was not so much in what Peter said, still less in any technique that he employed, but was rather the demonstration of the power of the Holy Spirit. His analysis of Peter's sermon on the Day of Pentecost is a pathetic travesty of the facts as recorded in the Acts of the Apostles. It is the result of a failure to understand the New Testament doctrine of the Holy Spirit and especially of the result of the fulness of the Spirit, in the ministry of men such as Peter, Paul or any of the great preachers who were raised up and used by God in subsequent ages. The idea of Peter's deliberately applying stimuli and manipulating his congregation and deciding when to 'hurl in' certain statements is, indeed, quite laughable" (p. 28).

With these sentiments I entirely agree. In all the characteristic Christian experiences, whether regeneration, sanctification, the baptism with the Holy Spirit, or divine healing, there is a unique and miraculous element present which can neither be produced nor explained by any kind of naturalistic theory. That element is the work of the Holy Spirit.

Dr. Lloyd-Jones goes on, however, to make certain admissions and to sound certain warnings in regard to brainwashing and evangelistic crusades. He says, "As I see the situation, we have to admit that wrong tendencies *can* develop and spread even among those who are sincerely desirous to spread the true Gospel. I think that in sheer honesty we must take note that some of these tendencies have crept into evangelical circles in the past" (p. 33). He then goes on to illustrate from the preaching of Jonathan Edwards, George Whitefield, and Charles G. Finney. What he has to say about this is, I feel sure, even more relevant to healing crusades. I think it cannot be denied that some things that have happened in divine healing meetings are more akin to brainwashing techniques than the miraculous workings of the Holy Spirit.

Dr. Lloyd-Jones strongly insists that if we really believe in the miracle-working power of the Holy Spirit we shall not put our confidence in psychological techniques. He says, "If what you desire is to produce psychological results, then, of necessity, you will have to employ the proper psychological techniques. But I am arguing that we are not to do so if we really believe in the work of the Holy Spirit. We are to present the truth, trusting to the Holy Spirit to apply it" (p. 39). He then makes a statement which is of paramount importance, because it shows the radical difference between preaching which is relying on psychological techniques, and preaching which is relying on the power of the Holy Spirit. He says, "In presenting the Christian Gospel we must never in the first place, make a *direct* approach either to the emotions or to the will. The emotions and the will should always be influenced through the mind. . . . Truth is intended to come to the mind. The normal course is for the emotions and the will to be affected by the truth after it has first entered and gripped the mind. . . . The approach to the emotions and the will should be *indirect*. Still less should we ever bring any *pressure* to bear either upon the emotions or the will. We are to plead with men but never to bring pressure. We are to 'beseech' but never to browbeat" (p. 39).

I think that here we have the basic difference between the spiritual and miraculous working of the Holy Spirit, and brainwashing. The Holy Spirit applies the truth to the mind. The brainwasher puts the mind out of action by pressurizing the emotions and the will. I think that in modern revivalism, with or without the healing emphasis, there is often confusion between the miraculous working of the Holy Spirit and a kind of psychological assault on the emotions and will of the people. I believe with all my heart that all who truly believe in New Testament revival and evangelism, and in New Testament miracles of conversion and healing, should altogether renounce and shun anything which savours of brainwashing.

There are two often-quoted passages of Scripture which are relevant to this theme. One is in the Old Testament: "Not by might, nor by power, but by my Spirit, saith the Lord" (Zechariah 4:6). The other is in the New Testament: "For the weapons of our warfare are not carnal, but mighty through God to the pulling down of strongholds" (2 Corinthians 10:4). Anything that savours of brainwashing comes under the censure of those statements. We must condemn brainwashing as a carnal might, a carnal power, a carnal weapon. Some might even denounce it as demonic; and the distinction between the carnal and the demonic is not very great! We cannot, in any degree, admit brainwashing into the sacred service of our God. If the Lord is not working with us "with signs following" (Mark 16:20), God forbid that we should shake ourselves like Samson when the "Lord was departed from him" (Judges 16:20). Rather let us humble ourselves before the Lord and seek His face for a new dedication and a new anointing until He Himself works His own miraculous works among us.

Second, Conservative Evangelical Criticism

Atheists and rationalistic theists are not alone in their bias against miracles in general and healing miracles in particular. Even within the world of conservative evangelicals there are many who raise the voice of criticism. Conservative evangelicals, of course, are totally committed to the concept of a miracle-working God. They express no disbelief in the miracles of the Bible, indeed they exult and glory in scriptural miracles. *What they object to is the prepetuation of miracles beyond the Apostolic era.*

It is generally recognized that there was a gradual falling off of miraculous phenomena, such as bodily healings, after the death of the Apostles and their immediate helpers. Those who advocate a doctrine of divine healing teach that this loss was due to a decline of spiritual power in the early church. For example, Dr. Evelyn Frost, in her work, *Christian Healing,* declares, "Even before the time of Constantine the beginnings of this decline are to be seen. The long peace corrupted the Church, and Cyprian indicated how through corruption the power of the Church had been weakened. With the decline from the first radiant fire of faith came a slow decline in the manifestation of the marks of the Spirit, both in the fruits and in the Charismata" (p. 181). In similar vein Dr. Leslie Weatherhead writes, "As Pentecost faded into past history, the fellowship began slowly to disintegrate, the faith of the Church in the power of the Holy Spirit, and that love for men which is derived only from love to God—diminished to such an extent that we note an insidious difference in the nature of the healing technique. Less is demanded from the healer. More is demanded from the patient. The onus is more and more put on the patient to have 'faith,' but the power to call forth faith is sadly lowered" (*Psychology, Religion and Healing*, p. 88).

Now there are many conservative evangelicals who will have none of this point of view. They claim that the loss of the phenomena is not due to man's unspirituality but to God's will. Man has not failed to appropriate miracles. It is God who has withdrawn them.

The change is explained in two different ways. One of these is plainly put by Dr. Henry Frost in the well-known book, *Miraculous Healing*. He claims that the miraculous phenomena were given to attest the ministry of Christ and the Apostles to the Jews. As that ministry was rejected the miracles ceased. "The miraculous acts increased to the degree the Jews accepted Christ, and decreased to the degree they rejected Him. When the Jewish Nation had finally rejected Christ, as also the Apostles, Stephen and Paul, miracles, including miracles of healing, almost ceased. What remained were isolated acts which corroborated the Apostolic authority and continued the witness to a living and loving Christ. When the time comes for a new offering of the Kingdom to Israel, miracle-working will be renewed (Revelation 11:3-6); and when the Kingdom has been established all of the prophetic promises concerning miracles, including healing, good health and long life, will be fulfilled. But now the Kingdom is not being offered to the Jews, for this is the Church Age. It is not, therefore, the age of miracles, except as God is pleased to manifest His power to individuals, in exceptional circumstances and for specific purposes" (p. 94).

This quotation raises one or two controversial issues which are beyond the scope of these lectures. One thing only need be said: Dr. Frost evidently does not eliminate miracles from the Church Age entirely. God may still work the occasional miracle, though the Church Age is not the age of miracles. Inasmuch as Dr. Frost allows for some miracles, his view is not a serious obstacle to us. In fact, the final chapter of his book describes a "great and notable miracle of healing" (pp. 118-125).

This view, which expresses the philosophy of many fundamentalists and dispensationalists, is less radical than that which is popular among more thorough-going Calvinists. According to the more radical view, New Testament miracles were given by God only to validate the New Testament revelation. They were part of the foundation-laying work of Christ and His Apostles, and when that work was done the miracles ceased and have not and cannot be restored. The miracles were an essential part of the Christian revelation. When that revelation was completed, and infallibly recorded in Holy Scripture by the inspiration of the Holy Spirit, the miraculous phenomena were withdrawn by God, never to be repeated. Dr. George Smeaton, for example, in his authoritative work, *The Doctrine of the Holy Spirit,* says, "The supernatural gifts in the New Testament Church were very abundant, but all culminated in the Apostles, who were the organs of Christ's revelation to the Church. . . . The cessation of these gifts, after they had served their purpose, was a significant fact. . . . These extraordinary gifts of the Spirit were no longer needed when the Canon of Scripture was closed. Up to that time they were an absolute necessity. They are now no longer so. Nor is the Church warranted to expect their restoration . . ." (pp. 139-140).

The greatest champion of this school of thought was Professor Benjamin B. Warfield of Princeton Theological Seminary. In 1917 Dr. Warfield gave a series of lectures at Columbia Theological Seminary, South Carolina, which was published in book form under the title, *Counterfeit Miracles*. This book became the standard work presenting the case against the continuance of miracles after the Apostolic Age, and in 1965 was reprinted under the title, *Miracles Yesterday and Today, Real and Counterfeit*.

In his first lecture Dr. Warfield lays down his basic hypothesis: "The Apostolic Church was characteristically a miracle-working Church. How long did this state of things continue? It was the characterizing peculiarity of specifically the Apostolic Church, and it belonged, therefore, exclusively to the Apostolic Age—although no doubt this designation may be taken with some latitude. These gifts were not the possession of the primitive Christian as such; nor for that matter of the Apostolic Church or the Apostolic Age for themselves: they were distinctively the authentication of the Apostles. They were part of the credentials of the Apostles as the authoritative agents of God in founding the Church. Their function thus confined them to distinctively the Apostolic Church, and they necessarily passed away with it" (p. 6). He qualifies this somewhat, later in the same lecture, by saying, "This does not mean, of course, that only the Apostles appear in the New Testament as working miracles, or that they alone are represented as recipients of the Charismata. But it does mean the Charismata belonged, in a true sense, to the Apostles, and constituted one of the signs of an Apostle" (p. 21). One further quotation may be given: "The abundant display of miracles in the Apostolic Church is the mark of the richness of the Apostolic Age in revelation; and when this revelation period closed, the period of miracle-working had closed also, as a mere matter of course" (p. 26).

This contention of Dr. Warfield is obviously a very serious matter indeed. If it is the truth of God, then it is the end of the entire quest for divine healing conceived as a miraculous act of Almighty God. The doctrine of divine healing is an error. Prayers for divine healing are vanity. The practice of the laying on of hands, or the anointing with oil, for sick people, is a meaningless piece of ritual. And the testimonies to miraculous healing are either delusion or fraud.

Dr. Warfield's ideas have great influence among some of the most intelligent and spiritually-minded evangelical Christians. Many Christian doctors accept and praise his opinions.

To those of us who believe in the concept of miraculous divine healing, however, these ideas are repugnant. Even though we greatly revere Dr. Warfield as a man of massive scholarship who did as much as, if not indeed more than, any man of his generation to defend the great foundational truths of the Christian faith against rationalistic attacks, we must part company with his negative opinions on miracles. This does not mean that his book is valueless. Indeed, it is extremely valuable. The bulk of it is concerned with a historical survey and appraisal of the quest for and claims of miracles throughout the Christian era. And whether we wholly agree with his opinions or not there is much that we can learn from it.

Time and space limit us to the making of three criticisms of Dr. Warfield's views.

First, Dr. Warfield offers no definite Bible statement that miraculous healings were exclusively for the validation of the Apostolic period and the Apostolic ministry. The reason for this is, of course, that no definite Bible statement can be found. What he offers us is, a hypothesis based upon his interpretation of certain Biblical facts. He presents us with three fundamental facts: 1. The Apostles were commissioned to a unique, foundation-laying work in the Church. 2. They were also the special media of revelation and inspiration for the New Testament. 3. Their ministry was accompanied and certified by miraculous phenomena. Now, we do not, and cannot, question these facts, as facts. But Dr. Warfield goes further. He links these three facts together in an exclusive and necessary relationship, and this is where we are compelled to join issue with his theory. He has no scriptural authority for doing this. It is pure assumption and speculation on his part.

The fact is, the miraculous healing ministry of Christ was not wholly or exclusively for the purpose of validating His Messiahship. That this was one motive for His miracles we have no doubt. But not the sole motive! *We must also give room for the motive of compassion.* For example in the narrative of the raising to life of the son of the widow of Nain, Luke says, "And when the Lord saw her, he had compassion upon her" (Luke 7:13).

Dr. E. H. Plumptre, writing in *Ellicott's Commentary*, says here, "Note, in this instance, as in so many others (e.g. Matthew 20:34; Mark 1:41), how our Lord's works of wonder spring *not from a distinct purpose to offer credentials of His mission, but from the outflow of His infinite sympathy with human suffering.*"

If the healing miracles of the Apostolic circle were, at bottom, a continuation in the post-resurrection period of the ministry of Christ (as the opening sentence of Acts implies), then surely we must allow room for the motive of compassion there, too. Once we allow this, we are bound to recognize that the healing miracles of Christ and the Apostles were not entirely bound up with the idea of validation. They were not merely proofs or credentials. *They were part and parcel of the gracious mission and ministry of redemption and salvation,* and not mere external adjuncts and guarantees.

Here, I think, we touch the basic difference between those Christians of Dr. Warfield's persuasion and ourselves. They limit miraculous healing to a "sign value," indeed, exclusively, a sign-value of the Apostolic period. We believe it is more than this. To some extent it was this, and is this. But also, it is an expression of the compassion of God for men and women in their need. It is an actual part of the salvation which is mediated to us through our Lord Jesus Christ.

Dr. Henry Frost allows some place for this. Though he does not place major emphasis on the compassion motive, but rather on the credential motive, yet he clearly states that "compassion was *a* motive in moving Christ to heal." Then he goes on to say, "Also, it must not be concluded that Christ had compassion upon the needy sons of men when He was on earth, but has no such compassion now that He is in Heaven. Christ is the eternal Son of God, and He is in His divine attributes, 'the same yesterday, and today, and forever' (Hebrews 13:8). If, therefore, He loved in the days of His flesh, He loves now; if He healed then, He will undoubtedly heal now" (p. 108).

Our second criticism of Dr. Warfield is this: It runs counter to several New Testament passages which imply the continuation of healing miracles so long as the Church exists. I shall especially take a look at John 14:12; 1 Corinthians 12:9; and James 5:14-16.

First, John 14:12: "Verily, verily, I say unto you, he that believeth on me, the works that I do shall he do also; and greater works than these shall he do; because I go unto my Father." Taken as it stands, this verse seems clearly to imply that the miraculous ministry of Christ would continue and increase after His return to the Father. There is no hint of a cessation of it, and, moreover, the phrase *"he that believeth on me"* cannot be limited exclusively to the Apostles and their helpers. It has a universal and age-long ring about it. Disciples of Dr. Warfield's opinion, however, do not accept this view of it. For example, Doctors V. E. Edmunds and C. G. Scorer, editing a pamphlet put out by the *Christian Medical Fellowship,* entitled *Some Thoughts on Faith-healing,* say: "Surely the 'greater works' referred to in John 14:12 are not intended to be comparable with the Messiah's signs of power. They are works of a spiritual rather than of a physical nature. . . .

"The greater works were not greater *miracles,* for who would dare to claim this? They were spiritual works in the hearts of men and women, consequent upon His resurrection, ascension, and gift of the Holy Spirit" (p. 32).

In reply we shall make the following observations: (a) The term *"works"* in reference to the ministry of Christ as used in John's Gospel, seems to have a special technical meaning. It seems clearly to refer to His miracles. The Greek word, of course, neither in its meaning or its usage, has this limited significance in the New Testament. But as used by John in reference to the ministry of Christ it does appear to have this special meaning. The following passages may be examined—John 5:20, "For the Father loveth the Son, and showeth him all things that he himself doeth; and he will show him greater works than these that ye may marvel." John 5:36, "But I have greater witness than that of John; for the works which the Father hath given me to finish, the same works that I do, bear witness of me, that the Father hath sent me." John 7:3, "His brethren, therefore, said unto him, Depart from here, and go into Judaea, that thy disciples also may see the works that thou doest." John 7:21, "Jesus answered, I have done one work, and ye all marvel." John 9:3, "Jesus answered, Neither hath this man sinned, nor his parents, but that the works of God should be made manifest in him." John 9:4, "I must work the works of him that sent me." John 10:25, "The works that I do in my Father's name, they bear witness of me." John 10:32, "Jesus answered them, Many good works have I shown you from my Father; for which of those works do ye stone me?" John 14:10, "The Father that dwelleth in me, he doeth the works." John 15:24, "If I had not done among them the works which no other man did, they had not had sin." These who interpret John 14:12 as purely "spiritual works" as distinct from physical miracles are going counter to the usage of the term "works" by John in reference to Christ.

(b) The alternative to thinking of the "greater works" as the spiritual works of conversion, is not necessarily the idea that any one believer can do greater physical miracles than Christ. Some people, of course, have imagined this. But this is a naive idea. A mere comparison between the Gospels and the Acts will show that even the Apostles themselves did not perform miracles greater either in quality or number than did Christ. The late Dr. William Temple in his well-known work, *Readings in St. John's Gospel*, goes to the heart of the problem in the following statement: "In scale, if not in quality, the works of Christ wrought through His disciples are greater than those wrought by Him in His earthly ministry. . . . The accomplishment of the *journey* to the Father means, among other things, that the Lord is no longer 'straightened' by the limitations of our mortal state; He is where God is, and that is everywhere. His works are no longer limited to Palestine but are diffused over the whole world" (Vol. 2, p. 235). *Ellicott's Commentary* also points this out: "The explanation of these greater works is not to be sought in the individual instances of miraculous power exercised by the Apostles, but in the whole work of the Church" (Vol. 6, p. 506). The *Moffatt Commentary* on John quotes Professor E. F. Scott: "The work of the Church is but a continuation under larger conditions, of the work of Christ Himself" (p. 308). In the days of His flesh our Lord was limited to His physical body in the little land of Palestine. From Pentecost He has a mystical body of believers through which He can work in a larger way. This, I believe, is the meaning of the expression "greater works than these shall ye do."

(c) There is a third observation to make: most commentators seem to have overlooked the fact that Christ not only said "greater works than these shall ye do," but also, "he that believeth on me, *the works that I do shall he do also.*" Even if we grant that the greater works are *spiritual,* it is evident that the "greater works" do not cancel out works of a miraculous nature. Jesus plainly stated that the works which He Himself had done would be continued through the Church in the post-ascension period, as well as the "greater works." Alford in his *Commentary on the Greek New Testament* sees this and states it plainly, though he views the greater works as "spiritual." He says, *"The works that Jesus did, His Apostles did, raising the dead, etc.—greater works than these they did, not in degree but in kind: spiritual works under the dispensation of the Spirit"* (Vol. 2, p. 770).

In my view, therefore, John 14:12 plainly implies that the miraculous ministry of Christ would continue after His ascension on a world-wide scale, not being limited to the Apostles but comprehending "him that believeth."

Second, 1 Corinthians 12:9. Here, Paul enumerates various *Charismata* which are bestowed by the Lord upon His Church, and manifested by the Holy Spirit. One of these "gifts" is *"the gift of healing."* Doctors Edmunds and Scorer write concerning the Charismata: "That some were miraculous there can be no doubt, and healing was probably one of these. But there is no suggestion that these were to continue indefinitely in the Church" (p. 62).

My own reading of 1 Corinthians 12 leads me to exactly the opposite conclusion: *there is no suggestion that the miraculous gifts were to be discontinued.* There is no hint in chapters 12 - 14 of the Corinthian letter that Paul is thinking only of the Corinthian situation or of the Apostolic Age. He is thinking of *the whole body of Christ* (see 12:12, 13). That body was not peculiar to the Apostolic Age, but belongs to the entire Christian era.

There are those who imagine that certain statements in chapter 13 imply that the miraculous gifts of the Spirit would cease at the close of the Apostolic period. "Whether there be prophecies they shall fail; whether there be tongues, they shall cease; whether there be knowledge, it shall vanish away. For we know in part, and we prophesy in part. But when that which is perfect is come, then that which is in part shall be done away"

(13:8-10). It is sometimes claimed that by *"that which is perfect"* Paul means the completed Canon of New Testament books, and that he is saying that the *Charismata* are only a temporary arrangement until that has come. Such an interpretation, however, has no warrant other than the desire to get rid of miraculous gifts. Alford says, "Chrysostom, and others, understand the words 'shall fail' and 'shall cease' of the time when, the faith everywhere dispersed, these gifts should be no longer needed. *But unquestionably the time alluded to is that of the coming of the Lord;* see v. 12, 'now we see through a glass darkly, but then face to face; now I know in part, but then shall I know even as also I am known'; this applies to all these three gifts which shall be superseded, not to the last only" (*Commentary on the Greek Testament,* Vol. 2, pp. 555-6).

Thus understood, these verses, which are supposed to limit the Charismata to the Apostolic era, actually give us ground for believing that they will abide in the Church right through the Church Age until the coming of the Lord.

There is another point of view concerning the "gifts of the Spirit" which also tries to relegate the miraculous phenomena to the Age of the Apostles. According to this view, there are two categories of *Charismata*: the *ordinary* Charismata, such as wisdom and knowledge, and the *extraordinary* Charismata, such as tongues, miracles, and healing. It is then claimed that the extraordinary gifts ceased with the Apostles, while the ordinary gifts will persist while the Church is on earth. This appears to be a very neat package, but it is completely arbitrary, having no foundation or support in holy Scripture.

There is nothing whatsoever in the Corinthian chapters to indicate that the "gifts of healing" (or any other Charismatic phenomenon) were to be limited to the Apostolic Age.

Third, James 5:14-15: "Is any sick among you? let him call for the elders of the church; and let them pray over him, anointing him with oil in the name of the Lord: and the prayer of faith shall save the sick, and the Lord shall raise him up; and if he have committed sins, they shall be forgiven him."

Various expedients are adopted to dispose of the applicability of this passage to the Church of all the Christian era. Dr. Henry Frost says "The instructions of James concerning healing were intended particularly for the Church in a condition of a large Jewish membership and at a time when it was emerging from Judaism and was spiritually undeveloped; and hence, they were not intended for the Church in its present Gentile condition and spiritual maturity" (*Miraculous Healing,* p. 68). This is pure assumption. Professor Warfield declares that the passage in James is "irrelevant," and contains "no promise of healing in a specifically miraculous manner." He strongly leans to the view that the anointing with oil referred to by James means "giving the sick man his medicine in the name of the Lord." He actually uses the expression *"rubbing him with oil* in the name of the Lord" (pp. 171-2).

Professor Rendle Short, in his book, *The Bible and Modern Medicine,* discusses the significance of the phrase *"anointing with oil"* by James. I quote his statement in full: "Two Greek words are translated 'anoint' in the Old and New Testaments, *aleipho* and *chrio.* When a symbolical or ceremonial anointing is meant, the latter word is used, as for instance 'God, thy God, hath anointed Thee with the oil of gladness,' and 'He that anointed us is God' (**Hebrews 1:9; 2 Corinthians 1:21**). When women anointed our Lord's feet with anointment, *aleipho* is used, and in secular Greek the rubbing of an athlete's limbs with oil was similarly expressed. In the Greek version of the Old Testament, *aleipho* occurs seventeen times, but is used in a ceremonial sense in only two of these. It may, therefore, well be that the anointing with oil in James, and in Mark 6:13 may be remedial rather than ceremonial. As we have seen, the ancients used external medicants freely. However, there is room for difference of opinion about this interpretation" (pp. 125-126).

I want to make three observations concerning this statement. First, Dr. Rendle Short offers this with hesitation and allows for differences of interpretation. Second, he seems to lean towards the view that the anointing with oil commanded by James for the sick was remedial, not symbolical. It seems to me, however, that this conclusion is not justified by his own argument. Though he refers to the Greek usage of oil for athletes he produces no relevant Old or New Testament passage in harmony with this practice, yet he jumps to the conclusion that James 5:14 and Mark 6:13 "may well be remedial rather than ceremonial." This is going beyond his own evidence. But, thirdly, I want to suggest that there is absolutely nothing in the New Testament usage of the words *aleipho* and *chrio* to lead to such a conclusion.

The word *chrio* occurs five times in the New Testament. Luke 4:18, "The Spirit of the Lord is upon me, because he hath anointed me to preach the gospel to the poor . . ."; Acts 10:27, "For of a truth against thy holy child Jesus, who thou hast anointed . . ."; Acts 10:38, "How God anointed Jesus of Nazareth with the Holy Spirit . . ."; 2 Corinthians 1:21, "Now he who stablisheth us with you in Christ, and hath anointed us, is God"; Hebrews 1:9, "God, even thy God, hath anointed thee with the oil of gladness above thy fellows." Four of these references apply to the Lord Jesus Christ; three of them specifically mean *His anointing with the Holy Spirit*. The other reference applies to Paul and his co-workers, and quite probably to all Spirit-filled Christians; again, it is the anointing with the Holy Spirit which is meant. This is quite different from what Dr. Short says. He says that *chrio* in the New Testament means a *"symbolical or ceremonial anointing."* Actually, it is only used of the *spiritual anointing* of *the Holy Spirit*. In the Old Testment it is used of "symbolical and ceremonial anointing" as in the case of Priests and Prophets and Kings, but in the New Testament never. *It always describes the spiritual reality.* Several authorities confirm this: Abbott Smith says it means "consecration to a sacred office." Cremer says, "In the New Testament *chrio* only occurs with reference to the Old Testament anointing, and as denoting a consecration and endowment for sacred service." Thus we see why James did not use *chrio*. He could not use *chrio* because his meaning would have been completely misunderstood. His readers would have imagined that he was referring to consecration to divine service, to the filling and anointing with the Holy Spirit, and this was foreign to his intention.

In regard to the word *aleipho*, W. E. Vine, in his *Expository Dictionary of New Testament Words*, says, "It is a general term for anointing of any kind, whether of physical refreshment after washing, or of the sick, or of a dead body" (p. 58).

There are eight occurrences of the word in the New Testament: (Luke 7:38, 46; John 11:2; 12:3; James 5:14; Matthew 6:17; Mark 6:13; Mark 16:1). The interesting fact is that not once is *aleipho* used in the sense of a medicine—unless we allow the usage in Mark 6:13 and James 5:14. I believe that these two passages are locked together in respect of this issue. If the *"anointing with oil"* was medicinal in one, it was so in the other. If it was symbolical or sacramental in one with a view to miraculous healing, it was so in the other. Personally, I find it impossible to believe that Christ sent forth His Apostles to use oil as a medicine for sick people. Only the bias against miraculous healing could lead to such an opinion. Alford says of Mark 6:13, "This oil was not used medicinally, but as a vehicle of healing power committed to them, a symbol of a deeper thing than the oil itself could accomplish" (*Greek Testament*, Vol. 1, p. 319). A more recent commentator, Dr. C. B. Cranfield, in the *Cambridge Greek Testament Commentary* on Mark, says, "Oil was used in the ancient world as a medicament, but its use by the Twelve was probably symbolic rather than medical in intention." He quotes Calvin as saying, "The oil was meant to be a visible token of spiritual grace, by which the healing that was administered by them was declared to proceed from the secret power of God" (p. 201). What was true of Mark 6:13 is also true of James 5:14. Alford declares, that "the anointing was not a mere human medium of cure, but had a sacramental character" (*Greek Testament*, Vol. 4, p. 326).

The question may be asked: Why then do James and Mark use the word *aleipho* and not the word *chrio*? To which I reply: they were precluded from using *chrio* because it was the special word for consecration to divine service. They could not have used this word without being grievously misunderstood. They took the more general word *aleipho* and used it to describe a symbolic or sacramental anointing. There is no thought of medicine, but only of miracle.

We come now to a third and final criticism of Dr. Warfield's views: he is forced, on his hypothesis, to approach the study of miracles of the post-Apostolic era with a closed mind. If miracles ceased with the Apostles then it is useless to look for them after that time, allowing, of course, for a reasonable transitional period.

Now, it is well-known that others have worked through the centuries of Church history and have found clear evidence of the continuation, not only of healing miracles, but of other Charismatic phenomena. George Jeffreys does this in his book, *Healing Rays*. He leans heavily on the work of A. J. Gordon in his book, *The Ministry of Healing*. Dr. Evelyn Frost, in her *Christian Healing,* and Dr. Leslie Weatherhead, in his *Religion, Psychology, & Healing,* both work through Christian history and find evidence of a continuation of miraculous healing ministry, in spite of the fact that the power to heal seriously declined in comparison with Apostolic days.

Dr. Warfield also works through history, especially medieval history and some 19th century groups, such as the Irvingite circle and Christian Science. There is this to be said: he works through the material more thoroughly and with much sounder judgment than many others. It would be a salutary exercise for all advocates of miraculous healing to read through his critical account. It serves as a grave warning to all of us concerning the danger of "over-belief" and superstition and wishful thinking. There is much to learn from his review of history. Yet the fact must be faced that his fundamental hypothesis precludes him from finding any real miracles of any kind in the Christian era.

A story is told in England of a professed atheist who was holding forth at the famous "Speaker's

Corner" in London's Hyde Park. He declared, "I have read the Bible through looking for eternal life, and I have not found it. *And what's more, I don't want to find it!*" To which an ironical voice called out, "That's why you ain't found it!" When you approach your search for anything with such a prejudice, it is almost a miracle if you find it! Is it possible that such a prejudice has blinded the great Dr. Warfield?

CHAPTER II

CLARIFYING THE TRUTH

Not only does the truth of divine healing have to be defended against those who deny and oppose it, but also it has to be preserved and protected from the corruptions and confusions which surround it.

Evangelicals and Pentecostals are not the only people to engage in a ministry of para-normal healing; that is, healing by methods other than the recognized natural techniques. Christian Science and Spiritualism give a large place to it. Dr. Louis Rose, a prominent British psychiatrist, has made a close study of the whole subject of faith healing and faith healers, and has presented his findings in a very readable book entitled *Faith Healing*. He says: "In England alone there are well over ten thousand psychic healers, including such diverse personalities as a Buddhist practitioner, a retired Army Officer, a blind man, a London grocer offering healing across a counter, a bacteriologist, a farmer, a yogi clergyman, a former cinema organist, reporters, and several civil servants. Others (one of whom claims an eighty per cent success rate) believe they are able to heal animals, and at least one has described the control of plant growth by blessing and cursing. Over two thousand of them are associated in a single body, and the demand for their services is such that some healing centers work on a mass-production basis with sufferers being attended to by adepts of varying degrees of skill."

Dr. Rose also goes on to show that there are great numbers of such "healers" in North

America, as well as in European countries, such as Switzerland, Germany and the Scandinavian countries (Norway, Sweden, Finland), and even in Roman Catholic Italy. He also states, "Even Communist China has its own school of para-normal therapy, with 'application of the thought of Mao-Tsi Tung' (i.e. dialectical materialism) having been claimed to prevent infection following severe burns" (*Faith Healing,* pp. 77-8, by Dr. Louis Rose).

Other religion systems, too, have a place for healing by faith. It is well-known that Billy Graham was once challenged by a Moslem leader in Africa to a kind of prayer contest to ascertain whose God was greater at miraculous healing. Moreover, there are people who seem to possess a "gift of healing" by nature. They have no religious beliefs or rites, yet they claim to impart physical healing by the touch of their hands.

In view of these things it is imperative that we clarify the truth of "divine healing in the atonement of Christ." It is commonly assumed by scientific investigators that the differences between these various practitioners are unimportant and trivial. They look for a common denominator, and find it in the emphasis on *"faith"* as the condition for the reception of healing. We shall try to show that there are differences between the "healing through the name of Jesus Christ" and purely "faith healing."

The issues are further complicated by confusions of thought, teaching and practice among Evangelicals and Pentecostals themselves. Any thoughtful person who has spent a few years in the circles where divine healing through the Atonement is taught cannot fail to be perplexed and even repelled by some of the ideas which are taught, and some of the methods practiced. Nor is this confined to the rank and file. Pastors and evangelists have sometimes been soures of fanaticism, and have led their followers into darkness and delusion. When ignorance speaks through a powerful personality, whether in the pulpit or over the mass communications media of our age, the damage to truth, as well as to people, can be enormous.

In our second lecture we shall take a look at some of the things that are causing confusion. We shall attempt to bring a little clarification to the confusions, and to suggest some remedies.

Misunderstanding the Nature of Faith

In the New Testament strong emphasis is laid upon faith as a condition of miraculous healing. Jesus said to the woman who touched the hem of His garment, "Thy faith hath made thee whole" (Matthew 9:22). To the father of a demon-possessed boy, He said, "If thou canst believe, all things are possible to him that believeth" (Mark 9:23). To Jairus He declared, "Fear not: believe only, and she shall be made whole" (Luke 8:50). When Peter was explaining the miraculous recovery of the lame man at the Beautiful Gate of the Temple, he said, "Christ's name through faith in his name hath made this man strong" (Acts 3:16), while James, in laying down the formula for anointing the sick with oil, said, "the prayer of faith shall save the sick" (James 5:16). There are many other declarations and allusions to the importance of faith for healing in the New Testament. There can be no doubt that, according to the New Testament, faith plays a vital part, both in the communication and appropriation of divine healing.

There is, however, great confusion in many minds in regard to faith. I want to draw attention to two of the most common of those confusions.

"Faith" is often confused with "suggestion" and "suggestibility." "Suggestibility" is a mental condition where ideas, either for good or for ill, can be easily received and responded to. Dr. Leslie Weatherhead defines "suggestion" as "the art of conveying an idea to the mind of another person in such a way as to make him entirely accept it, apart altogether from the *evidence* of his reason" (*Psychology, Religion and Healing,* p. 118). "Suggestibility" can be induced and manipulated either by oneself or by others, and this technique is called "suggestion." If "suggestion" is performed by oneself it is called auto-suggestion; if performed by others, it is called hetero-suggestion.

"Suggestion" is used in various areas of life today. It is a much used technique in the medical profession. To psychiatrists and psychologists, of course, it is a number one tool. But also it is greatly used in the political world, and is a potent weapon in the field of "psychological warfare." Perhaps it is the advertising industry which makes the greatest and most successful use of it. It is the power of suggestion that makes the super-salesman.

Those who study the healing miracles of Christ in the modern scientific spirit sometimes claim that Jesus healed by the power of suggestion. He cast out from men's minds the negative thoughts which caused sickness and disease, and imparted new and positive ideas that produced health and well-being.

Now even if this is all that Christ did, we must not despise it. Even if this is all that happens today when men pray for the sick "in the name of Jesus," we must not despise it. But if this is all that we mean by divine healing, then there is nothing unique (at least in this respect) about Christ. It means that He is no different from the healers of other religions, no different from the ten thousand psychic healers of England, no different from those healers who have no religious affiliations whatsoever.

The New Testament, however, presents Jesus as healing the sick, not by the natural, psychological force of "suggestion," but by the supernatural power of God. It would be fatal for us to give way on such an issue. Moreover, the "faith" for which Christ asked cannot be reduced to the mental condition of "suggestibility." *There are elements in true "faith" which are missing from "suggestibility."* Dr. Weatherhead defines Christian faith as "the response of the whole man, thinking, feeling and willing, to the impact of God in Christ, by which man comes into a conscious, personal relationship with God" (*Psychology, Religion and Healing,* p. 429). Dr. W. B. Selbie has a similar statement in his book, *The Psychology of Religion*: "We need not hesitate to admit the power of suggestion or to use it for religious purposes, but we cannot escape the conclusion that it finds its sphere of operation in the subconscious and instinctive side of our nature. It needs therefore to be supplemented by some national and reflective process before it can be fully effective" (p. 300).

We see here, therefore, a clear distinction between "suggestibility" and "faith." "Suggestibility" is almost wholly an emotional response to a powerful stimulus, with the reasoning and volitional sides of our nature inhibited. "Faith" is the response of the whole person, emotion, reason, and will. In *suggestion* the person is acted upon, compelled, by a strong personality or mass contagion. In *faith* the whole person acts, intelligently and deliberately, towards another person or towards a certain end.

What does this mean in practice for those of us who believe that Jesus Christ heals the sick today? It means that we cannot and must not be content to use the techniques of suggestion in our ministry to the sick. We are not amateur psychiatrists. And we are not professional crowd-compellers. We are not even healing practitioners. We are only true to our calling and ministry as we exert all our influence to help men and women into a vital, living faith in Christ Himself. The Apostle Peter, anxious to divert attention from himself to Christ, cried, "His name, through faith in his name, has made this man whole." In other words, it was the living Lord Jesus Himself who performed the miracle. What Peter did was to encourage the lame man to put his expectation, his hope, his trust, his confidence, in the Lord Jesus Christ. This was Apostolic healing. This is true Christian healing. And this is a rational and volitional process. It is not irrational. It is not superstition. It is not mass psychology. It is not focusing attention blindly upon religious relics or powerful personalities. It is understanding, intelligent, conscious faith fixed in Jesus Christ Himself.

"Faith is also confused with 'Positive Thinking.'" It is not my wish to pour scorn on the pragmatic values of "positive thinking." There can be no doubt that what is called a "positive outlook" on life, embracing cheerfulness, optimism and confidence, is far better than the opposite attitude of negativeness, fear, pessimism and gloom.

We cannot, however, equate positive thinking with Christian faith. Some people are born with a positive outlook. Religion or no religion, it is their nature to think positively. Other people have developed a positive outlook by cultivation, yet are far from holding the Christian faith as set forth in the New Testament. Two of the movements which deviate from true Christianity and yet make great use of the positive-thinking concept are *Christian Science* and *Father-Divinism*. Though poles apart in some ways their movements have much in common. Sara Harris, in her biography of Father Divine, says, "Christian Science and Divinism have some points in common. As a matter of fact, many of the tenets of Father Divine are Christian Science which he has adopted and exaggerated" (*The Incredible Father Divine*, p. 226). The positive attitude is one of these things in common.

Dr. J. Oswald Sanders shows that Christian Science also has strong affinities with New Thought, Unity and Theosophy. He says, "All four are pantheistic in their philosophy." He also quotes Pandita Ramabai, who became an evangelical Christian, as stating, during a visit to New York, that on meeting up with Christian Science in India she "recognized it as being the same philosophy that has been taught among her people for four thousand years" (*Heresies and Cults*, pp. 56 and 43). She refers of course to the Oriental religion of Hinduism.

Professor William James devotes a chapter of his famous book, *The Varieties of Religious Experience*, to what he calls *"The Religion of Healthy-mindedness or Mind-cure."* He shows that while its strongest advocate was Mrs. Eddy of Christian Science, yet it was also the philosophy of Walt Whitman, Emerson and Ralph Waldo Trine. I give one important quotation from James: "One of the doctrinal sources of Mind-cure is the four Gospels; another is Emersonianism or New England transcendentalism; another is Berkleyian idealism; another is Spiritism, with its messages of 'law and progress' and 'development'; another, the optimistic popular science of evolutionism . . .

and finally, Hinduism has contributed a strain" (p. 94).

I believe that this kind of philosophy has nothing to do with New Testament Christianity. If we let these ideas into our thinking they will only bring confusion and corruption. Furthermore, this healthy-minded attitude is far from being the cure-all that its advocates claim. I knew a Christian woman who tried to apply "positive faith" to a ghastly breast cancer of which she was unmistakably dying. Refusing to have medical treatment, she claimed that she was healed in the "invisible" —this in spite of dreadful pain and other outward signs to the contrary. She was never healed, and finally died of the complaint. I am sure that she is not the only victim of this type of thinking. There are many, not only in Christian Science and kindred cults, but among Evangelicals and Pentecostals, who imagine that "faith" means that they should abandon all means and helps, and indeed all good sense. I am convinced that this is a corruption and confusion of the truth.

Wrong or Suspicious Methods

It is a well-known fact that when large crowds of people gather together in a highly-charged emotional atmosphere, many people can be manipulated and influenced to do the most unusual and, indeed, the most irrational, things. People who would not normally or easily be influenced as responsible thinking individuals can lose their self-control in a crowd, and get carried along by a floodtide of powerful feeling which has been let loose or worked up among the crowd. This was the sort of thing which operated in the old days of lynch-law. We have become familiar with it currently in civil rights demonstrations and mob-violence. The majority of people who engage in these exhibitions are being driven along by the contagion of mass feeling which has been engineered by some powerful personality or mastermind. For thousands of years politicians and dictators have been familiar with the laws of crowd-psychology. They may not have studied the subject in a university! Nevertheless, they were experimentally acquainted with its workings. In modern times we have witnessed what mass psychology can do in the political and nationalist areas in the phenomenon of the political dictators, such as Hitler, Mussolini and Stalin. Dr. Arturo Castiglioni describes these men as "wizards" who use "a suggestion that kills" (*Adventures of the Mind*, pp. 333-369).

How does this apply to evangelism and the divine healing crusade? Since the days of Jesus and the Apostles people have often gathered together in crowds to hear the Gospel. Whitefield, Wesley, Moody, to mention but a few, frequently addressed large audiences, while "mass evangelism" has become a feature of modern times. It would be foolish and unbiblical to ban preaching to crowds because evil and unscrupulous men have exploited crowds in their own interests. Yet one thing seems to me self-evident: namely, neither Jesus nor His Apostles deliberately manipulated crowds of people by mass psychology to get "results." Sometimes, indeed, Jesus repulsed the crowds, rather than manipulated them, by the concepts of truth which He proclaimed and taught. It seems to me that here we have a guideline in dealing with crowds, if we want to avoid the dangers attending mass meetings. The deliberate use of mass psychology, either for evangelism or healing, is part of "the wisdom of this world" which we ought to refrain from in the higher interests of the Kingdom of God. As Dr. Martin Lloyd-Jones says (in his excellent little book, *Conversions: Psychological and Spiritual*, p. 40): "It is surely our business to avoid anything which produces a merely *psychological* condition rather than a *spiritual* condition."

Dr. Louis Rose draws attention to two of the dangers associated with healing meetings. The first of these is *hypnosis*. He says, "There is no doubt that . . . hypnotism can produce startling short-term effects with a sprinkling of more enduring ones, and these must account for a proportion of the faith-healer's apparent success. It is not suggested that the healer frequently and consciously employs hypnotic techniques. . . . But in any activity which takes place in an emotionally-charged atmosphere (and particularly when mass meetings are involved) cases of self-induced hypnosis may well occur. And when they do occur they can lead to superficially impressive results" (*Faith Healing,* p. 129).

I am quite sure that I have seen this kind of thing many times in healing crusades. I have seen people in the healing line, when the evangelist got within a few yards of them, as he went around in his ministry of prayer, fall headlong to the ground in a sort of coma. Rarely did any sort of permanent benefit result, and I believe that in most cases it was nothing more than self-induced hypnosis.

The other danger which Dr. Rose mentions is "*role-playing.*" He means by this, the tendency to act in a certain pattern in certain situations and circumstances. He mentions "the conduct of people at a wedding, a funeral, a job interview, in courtship or in battle." "One wishes to conform: one puts on an act." In "role-playing" a person is not in a state of hypnosis, and there is no loss of the critical faculty. He is, however, motivated to do the expected thing and fulfill a particular role.

We can easily see how this tends to operate in modern mass evangelism. In the main, the meetings of these mass crusades follow a set pattern which has become almost a ritual, even to the singing of a particular hymn during the appeal. Everything is planned to condition people to the act of "coming forward" for counselling at the close of the sermon. Now we know that the motive is pure; it is the spiritual motive of bringing people to Christ. However, there is grave danger that it becomes little more than "role-playing." People have the feeling that they should act in a given manner in these crusades, and tend to "come forward," not because they are genuinely seeking Christ, but are merely playing a role, like people at a funeral or a wedding. The large number of crusade converts who never make the grade (so unlike Peter's converts on the Day of Pentecost who not only "received the Word" and "were baptized," but "continued steadfastly in the apostles' doctrine and fellowship, and in breaking of bread, and in prayers") is an indication that there is something not right about this method, and maybe this theory of "role-playing" explains it.

The danger is equally present in the healing crusade, among the people who are prayed for at these meetings, especially if they are dealt with on a platform under powerful lights by a strong personality, with thousands of eyes gazing at them. They feel that they are expected to make certain responses, to act in a particular manner, to play a role. In consequence, there are similar anomalies in the number of healings as in the number of conversions at the ordinary evangelistic meeting. It is well-known that people sometimes testify to healing, of blindness or deafness maybe, during the healing demonstration, and even act as if they are healed, yet within days, sometimes within hours, it becomes absolutely clear that they are not really healed at all. They have simply been playing a role. There is no conscious or deliberative fraud, either upon the part of the patient or the evangelist. It is the method of the mass demonstration that produces the psychological phenomenon of role-playing. An uncle of mine, a cripple with rheumatoid arthritis, for nearly thirty years, was a godly man, and was often prayed for by pastors and evangelists. On one occasion he was taken to a large healing crusade conducted by a world-famous British evangelist. During this meeting my uncle, in response to suggestions from the evangelist, discarded his sticks and walked, even ran, several times around the great auditorium. To the crowd present a miracle had occurred, but the next day my uncle was in great agony of body. He had not been healed, and he never was healed to the day of his death. Neither he nor the godly evangelist was guilty of fraud. I believe it was a case of role-playing.

SENSATIONAL PUBLICITY AND EXAGGERATED REPORTING

One of the problems calling for urgent attention in the divine healing movement relates to the reporting of miraculous cures. The disparity between the reports of those who believe in healing miracles and of those who are skeptical is absolutely staggering. Judging by the claims of the various people who conduct large healing crusades there must be many, many thousands of people in the world today who have been miraculously healed. Yet some investigators claim that they cannot find evidence for one single miracle of healing, even after years of careful search and examination.

Let me quote from one or two sources. One British evangelist who is continually employed in healing crusades in various parts of the world makes the following statement, in a pamphlet, *The Significance of the Divine Healing Ministry*: "It is certainly true to say that a great number of these coming to be prayed for in our meetings are suffering from an incurable disease, at least by medical standards. I would frankly acknowledge that many of these come to the services seeking divine healing as their final resort. Yet these are the very people whom we are seeing healed by the power of God. During the last few years (he writes in 1963) we have prayed for perhaps 20,000 in person, and we have received hundreds of unsolicited testimonies of healing from almost every kind of sickness and disease. I have seen God heal 'incurable' cancers and tuberculosis, arthritis, poliomyletis, heart disease, the crippled, the paralyzed and the insane. I have even seen the Lord give sight to eyes born blind; many who were deaf have had their hearing restored, and even the deaf and dumb have begun to hear and speak" (p. 9, Brian Williams). Other healing evangelists make similar claims, and such miracles are being reported continually in a variety of magazines and pamphlets.

On the other side consider the following. Dr. Louis Rose, to whose book, *Faith Healing,* I have already referred several times, describes his careful and critical investigation of alleged miracle cures covering a period of fifteen years, and at the end of his book makes the following declaration, which is as startling negatively as the claims of the healing practitioners are startling positively: "After well over fifteen years of work I have yet to find one 'miracle cure'; and without that (or alternately, massive statistics which others must provide) I cannot be convinced of what is commonly termed faith healing. But my mind is not closed to new evidence, and I remain interested in examining any claims of unorthodox cures with hard facts to support them" (*Faith Healing,* p. 172). In harmony with this is the conclusion of Doctors Edmunds and Scorer (both Evangelical Christians) in their book, *Some Thoughts on Faith-Healing* (sponsored by the Christian Medical Fellowship): "The fact is that, from an examination of the results of the work of faith-healers and of healing missions, which are recorded in current writings, the impression is gained that little which can honestly be said to be truly miraculous occurs today. Disappointing as it may seem, the facts do not warrant the rather sweeping assertions and self-advertisements of many of the healing practioners" (p. 64).

These doctors make a special reference to Pentecostals in regard to this matter. They say, "Among Protestants, the Pentecostalists make the most definite claim that all the special Apostolic gifts are still in operation, and state that they continually have evidence of the efficacy of the gift of healing. It is difficult, however, to secure accurate case records. Of those which have been collected, it is not certain that clear cases of incurable organic disease are among the successes reported" (pp. 52-53).

This gap between the two approaches to miraculous healing is absolutely staggering. How does it come about that some people find evidences of miracles in hundreds, indeed, thousands, of cases, while others find no evidence at all? Both sides tend to blame each other's methods, if not each other's motives. The advocates of healing accuse the medical investigators as biased and critical, while the doctors think the healing advocates are credulous and motivated by wishful thinking.

As a Pentecostal, I feel keenly the charges

brought against Pentecostal evangelists in regard to this matter, and I am convinced that the whole Pentecostal movement should do something to put things straight.

It is well-known that the Roman Catholic Church has an authorized bureau which examines alleged miracles of healing at Lourdes. According to Dr. Louis Rose, this bureau, which is made up of Roman Catholic physicians, "examines the thousands of claims of miraculous cures made every year: of these claims about fifty pass a first screening, six a second one, and perhaps a single case is left at the end to be accepted as a probable miracle" (page 92). Dr. Leslie Weatherhead says that "In 1948 more than two million pilgrims visited the shrine. Fifteen thousand were patients, but only one is expected to be an official 'cure,' although eighty-three dossiers were retained for further examination" (p. 153). Professor Warfield quotes Monsignor R. H. Benson, a Catholic authority, as saying, "During the twenty years from 1888 to 1907, inclusive, the whole number of recorded cures was 2,665, which yields a yearly average of about 133" (p. 107).

This Roman Catholic bureau has certain fixed standards by which to test alleged cures. To quote Dr. Rose again, "It must, for instance, be noteworthy and, by Catholic standards, edifying and reasonable. It is usually instantaneous, though some miracles developing over a period of time have been accepted: it occurs in answer to prayer, though not necessarily the prayers of the person concerned; and where this is relevant, it must be persistent in its effects. Finally, it cannot be accepted as a miracle until there is an overwhelming weight of evidence that the event did in fact occur and that any natural explanation is supremely unlikely" (p. 87).

Misunderstanding and Misapplication of Holy Scripture

There are, even among Evangelical and Pentecostal Christians, a number of extreme and fanatical ideas which often result in much unhappiness. These ideas are frequently due to a misinterpretation of Bible revelation. It is imagined that the fanatical idea is taught in Scripture, whereas it is based upon a misunderstanding or misapplication of what the Bible teaches. Error is often distorted truth. There are many such distortions, but we shall mention only a few.

The idea that every sickness must be the direct result of personal sin. I shall not take time to quote any writer who advocates such an opinion. It is extremely doubtful if any responsible author could be found who would make such an assertion in print, though it is sometimes stated or implied from the pulpit. As a general rule this idea exists in the minds of uninformed believers. They are often haunted by a feeling that their sickness, especially if it persists for a long time in spite of prayer and anointing with oil, must be a divine judgment because of personal sin.

We need to understand that this notion is a distortion of a Bible revelation. The Bible does reveal that "the law of sin and death" operates in this world (Romans 8:2), and that "death" has entered into human experience as the consequence of sin (Romans 5:12). "Death" is a comprehensive term for all forms of sickness and weakness which lead finally to the separation of soul from body. It includes *disease* as well as *decease*. But the Bible does not teach that when an individual is sick it is always the consequence of his personal transgression and wrong. While, on the one hand, Scripture refers to individuals, such as Herod Agrippa

(Acts 12:20-25), who suffered because of their personal sins, it also, on the other hand, refers to individuals whose lives were patterns of holiness and faith, yet they were sick. We may mention Job, Elisha, Epaphroditus, Timothy, and Paul himself.

Derek J. Prime (in an article on "Sin and Sickness" in the symposium, *The Question of Healing,* edited by Gilbert W. Kirby), draws attention to the statement in James 5:15, 16: "And the prayer of faith shall save the sick, and the Lord shall raise him up: *and if he have committed sins, they shall be forgiven him.*" He goes on to say, "The important words, in our present consideration, are the two words *'and if.'* It is not envisaged that a man will be entirely without sins, but it is envisaged that he may suffer without there being specific sins which have merited illness. On the other hand, particular sins may have brought illness as a chastisement upon the individual, and genuine repentance and the prayer of faith will bring God's forgiveness and healing" (p. 80).

"*If* we have committed sin," certainly we must confess it and renounce it as a step towards our healing. But we may be walking in the pure light of holiness and in radiant fellowship with God, yet still be sick. Many people have been brought into great darkness of mind through not recognizing this. The Bible certainly teaches that sickness and death in a general sense have come into the world as the result of original sin. But the Bible does not teach that every case of sickness and disease is due to the sufferer's own sin.

The view that a person must be a fully committed Christian before the Lord will heal him. Again this is a notion which exists vaguely in people's minds. And sometimes it is sown in people's minds by suggestions from the pulpit. It is stated that the salvation of the soul is the first priority, that people must first come to Christ and commit themselves to Christ, *then* they can be healed. Unfortunately, this comes dangerously near to becoming a bargaining counter: "If you will first profess conversion, I will pray for the healing of your body."

There is no foundation for this concept in the New Testament. In the Synoptic Gospels nothing is more outstanding than the free, generous and wholesale manner in which Jesus healed the sick. When "He healed many that were sick of divers diseases, and cast out many devils" (Mark 1:34), He did not first insist that "they take up their cross and follow Him." When He cleansed the leper He did not first make a bargain with him. He healed out of compassion, and compassion does not strike bargains or demand conditions. In the Gospel according to John this is strikingly seen in several of the great miracles which John especially describes. When He approached the impotent man at Bethesda, He laid down only one condition, *"Wilt thou be made whole?"* (John 5:6). Not, "If you are willing to follow me, then I will heal you." He did not even tell the man who He was!

It is altogether wrong to use the promise of healing as a bait to induce people to make a profession of surrender to Christ. Jesus did not do it. Neither did the Apostles. As I read the Gospels I get the impression that Christ healed far many more people than actually had spiritual perception and saving faith. I get the same impression from the Acts of the Apostles. I do not get the impression that the multitudes who were healed through the ministry of Peter in Jerusalem, or of Philip in Samaria, or of Paul in Ephesus, necessarily became Christians either before or after their physical healing. I think that one can have faith to receive healing without having real saving faith. And one can have real saving faith without having faith to receive physical healing.

The idea that there is a necessary incompatibility between faith in God for healing and the use of means towards healing. I suppose that this is one of the most common confusions existing in the minds of those evangelical Christians who believe in divine healing. They imagine that it is inconsistent to pray for divine healing while still visiting a doctor or takng medicine. In spite of the fact that some of the most-used evangelists in the ministry of healing, such as George Jeffreys, have repeatedly denounced this notion, and even an entire denomination like the Pentecostal Holiness Church has also exposed the falsity of it in its *Manual of Discipline,* many people continue to refuse medical and psychiatric help under the mistaken idea that it is contrary to faith in God.

I have devoted an entire chapter of my book, *Sickness, Health and God,* to this subject. I do not believe that there is any real foundation in Holy Scripture for this belief. Not only is there no clear Biblical passage which teaches it, but also there are some passages which clearly imply that the use of means for bodily sickness is not prohibited. Isaiah's "lump of figs" for the healing of Hezekiah's boil will readily come to mind (2 Kings 20:7). Also Paul's exhortation to Timothy to "take a little wine for thy stomach's sake and for thine often infirmities" (1 Tim. 5:23), a passage which shows that sickness was neither unknown among the Apostles, nor was the use of means despised or prohibited.

In our Lord's ministry there are one or two striking examples. One of these is the use of clay and saliva in the healing of the man who had been born blind (John 9:6). Why did Jesus use these helps? There are two chief opinions. Barclay shows that the use of saliva as a simple remedy was common in those times, though repulsive to modern people. He gives a number of examples from ancient Roman history and says, "The fact is that Jesus took the methods and customs of His time and used them" (*Daily Study Bible*: John). Leslie Weatherhead adopts the view that Jesus used the spittle and clay, and the act of washing in Siloam, as a *suggestive therapy* to help the man to believe (*Psychology, Religion and Healing*).

Now, it is impossible to believe that the spittle and clay, whether used as a simple ointment, or a psychological technique, can account for the miracle. Only a supernatural act of Christ can explain it. These means were not even necessary to Christ. Usually He worked without means. It was not even necessary for Him to *touch* people, or to use oil, or even to speak directly to them, or be physically near to them. Sometimes He healed via these methods, sometimes without them. But the fact remains that in this miracle He used means. Whether we interpret the means as a simple, popular remedy, or as a psychological technique to create confidence and motivate to positive action, the fact remains that Jesus employed means, but blessed and used those means infinitely beyond their power.

I, too, believe that Christ can heal us miraculously. But I also believe we ought not to despise means. We ought to pray for doctors and nurses and all who are seeking to cooperate with the law of healing which God has placed in nature. It is not a condition of miraculous healing that we should first of all abandon the use of means. It is not a sign of unbelief, nor is it any mark of inconsistency, to be using medical help and praying for a miracle at the same time. There are not two Gods—a God of nature and a God of miracles. It is the self-same God who works in both realms. Whether He heals by nature or by a miracle, it is the same God who heals. We ought to bow to His sovereignty, and gladly accept our healing from His gracious hands, whatever the channel through which it comes.

The teaching that it is always the will of God to heal the sick, and that there can be no exceptions to this principle. Some well-known evangelists take a strong line on this question. They insist that Jesus always healed all who were sick in the multitudes of people who thronged about Him, and that this is the normal pattern for us.

I would not for a moment wish to minimize the vastness of our Lord's public ministry, but I would like to make two comments about the deduction drawn from it. First, though Jesus *on certain occasions* healed all who were present, it is extremely doubtful that He therefore always healed all, and that, moreover, there were no sick people in the Apostolic churches. This would be going beyond all the evidence. The most that we can say is that on *some occasions* Jesus healed all. We cannot infer that He always did so. And we cannot infer that there were no sick people in the Apostolic church, or even in the Apostolic company. We know that sickness did occur among New Testament saints and ministers, even as it did among Old Testament worthies such as Elisha, "who fell sick of the sickness of which he died" (2 Kings 13:14). Epaphroditus fell sick (Philippians 2:25-30). Timothy had many infirmities (1 Timothy 5:23). Paul left Trophimus sick at Miletus (2 Timothy 4:20). Paul himself suffered from a "thorn in the flesh" (2 Corinthians 12:7), which I believe to have been some physical disability.

CLARIFYING THE TRUTH 61

Nowhere is it suggested that these saints were sick because they were living in sin or lacked faith to appropriate healing. The fact is stated without any implication that it was sub-normal or out of pattern.

The second comment I would make is this: the theology which claims that it is always the will of God to heal the sick, is completely unrealistic and impractical in its outworking. During my thirty-five years as a Pentecostal pastor I have attended many healing campaigns conducted by various evangelists. I have seen many thousands of sick people prayed for in these meetings. In comparison, how few have been healed! If it is the Lord's will to heal all, why are so few healed in these meetings? To say that all the unhealed have no faith is arbitrary. Often as I have looked at them their eyes have been filled with expectation and hope, and as I read the Gospels I get the impression that Christ asked for no more than this degree of faith. If they had no faith they would not be in the healing line. And if the evangelist had no faith, it is difficult to think that he would be preaching and practicing divine healing. Yet so few receive healing! The paucity of results should make us seriously question the thesis that it is always God's will to heal the sick.

This idea can be a dangerous notion to get in the mind. If it doesn't work out—and how frequently it doesn't work out—a sensitive soul can be driven to morbid introspection and condemnation and guilt. Self-blame and self-despair can eat out the heart and destroy faith altogether.

James tells us that "we ought to say, If the Lord will, we shall live, and do this or that" (James 4:15). That is a principle which applies more widely than what James has specifically in mind. *"If the Lord will"* should be the context of all of our living and working, and all of our praying. Some people poke fun at the thought of praying, "Thy will be done." I am not one of those people. I believe that it is the highest form of faith to rest in the will of God.

Chapter III

INTERPRETING THE TRUTH

I should like to repeat here a statement which I made early in our first lecture: "In my opinion, one of the most necessary and urgent tasks, not only for the Pentecostal Holiness Church but for all earnest believers in the doctrine of divine healing, is to come to grips with the problem of healing in the Atonement, and to formulate a truly Biblical statement of what is meant by it." I also expressed the opinion in the first lecture that it is providential that our Article of Faith on divine healing says so little about the subject. It reads, *"We believe in divine healing as in the atonement."* This means that ministers and members of the Pentecostal Holiness Church are committed to two concepts. They are committed to the concept of divine healing, and they are committed to the concept of divine healing as in the atonement. But they are not essentially committed to the various interpretations and applications of these concepts which are common and popular in these days. In my opinion some people interpret and apply divine healing truth in foolish and fanatical ways, sometimes in questionable and dangerous ways. The Pentecostal Holiness Church is not committed to these interpretations and applications. We are committed to the truth of our twelfth Article of Faith, but we are free to examine, evaluate, accept or reject all of the many interpretations and applications of the truth which swarm around us. We ought to be grateful, both to God and our forebears, for this. They have bequeathed to us a precious heritage of Bible truth, but it is couched in such terms as to allow a large measure of freedom in thought and in expression. There is, however, another consequence. With freedom comes responsibility. We must beware of thinking that all interpretations and applications of divine healing truth are equally valid. I believe the day has come when we ought to separate "the precious from the vile"; to thoroughly examine, especially, the concept of "healing as in the atonement," perhaps, even, to set up a commission of competent persons to do this, who can produce a careful and authoritative statement of what the Pentecostal Holiness Church means by "divine healing as in the atonement."

In the present lecture I shall do four things: first I shall review the development of the concept of "healing as in the atonement"; second, I shall describe some criticisms which have been levelled at the concept; third, I shall discuss the various opinions of Pentecostals; and fourth, I shall suggest, tentatively, an interpretation of the concept which could command our respect.

A Brief Review of the Concept of "Divine Healing as in the Atonement"

As far as I know, the concept of healing in the atonement was unheard of until the latter half of the 19th century. At any rate I have found no references to it, apart from a doubtful one in Edward Irving, who lived a little earlier in that century. This does not mean, of course, that the truth of divine healing was unknown prior to the latter part of the 19th century. It was the emphasis on healing in the atonement that was new.

About that time several leading Bible teachers began to lay emphasis on three particular passages of Scripture—Isaiah 53:4, 5; Matthew 8:16,

17; 1 Peter 2:24. From these passages they drew the conclusion that when Christ died upon the cross, He not only suffered vicariously and as a substitute for humanity's sins, but also for humanity's sickness and disease.

For example, here is a statement from the well-known book, *The Gospel of Healing,* by Dr. A. B. Simpson, first published in 1888. Quoting Isaiah 53:4, 5, "Surely he hath borne our griefs, and carried our sorrows . . . and with his stripes we are healed," Dr. Simpson says: "This is the great Evangelical vision, the Gospel in the Old Testament, the very mirror of the Coming Redeemer. And here in the front of it prefaced by a great Amen—the only 'surely' in the chapter—is the promise of healing; the very strongest possible statement of complete redemption from pain and sickness by His life and death." He goes on to say, "The translation in our English version does very imperfect justice to the force of the original. The translation in Matthew 8:17 is much better: 'Himself took our infirmities, and bore our sicknesses.'" Dr. Simpson goes on to show that the two words translated "borne" and "carried" in Isaiah 53:4 "denote not mere sympathy, but actual substitution and the removal utterly of the thing borne," and declares, "Therefore, as He has borne our sins, Jesus Christ has also *borne away and carried off* sicknesses; yes, and even our pains, so that abiding in Him, we may be fully delivered from both sickness and pain" (pp. 11, 12). Later in his book, Dr. Simpson develops his thesis more theologically. He writes, "But redemption finds its centre *in the Cross* of our Lord Jesus Christ, and there we look for the fundamental principle of divine healing, which rests on the atoning sacrifice. This necessarily follows from the first principle we have stated. If sickness be the result of the Fall it *must be included in the atonement* of Christ, which reaches as far as the curse is found" (p. 31). In the same context, Dr. Simpson quotes 1 Peter 2:24, "His own self bore our sins in his own body on the tree . . . by whose stripes ye were healed," and declares, "In His own body He has borne all *our bodily liabilities* for sin, and our bodies are set free. That one cruel 'stripe' of His—for the word is singular—summed up in it all the aches and pains of a suffering world, and there is no longer any need that we should suffer what He has sufficiently borne. Thus our healing becomes a great redemption right, which we simply claim as our purchased inheritance through the blood of His Cross" (p. 32).

Another revered teacher who used similar language was Dr. Andrew Murray, in his book, *Divine Healing*: "It is not said only that the Lord's righteous Servant had borne our sins (v. 12), but also that He has borne our sickness (v. 4, R.V. margin). Thus His bearing our sicknesses forms an integral part of the Redeemer's work as well as bearing our sins. Although Himself without sin He has borne our sins, and He has done as much for our sicknesses—as soon as a sick believer understands the purport of the words, Jesus has borne my sins, he does not fear to say also: I need no longer bear my sins, they are upon me no longer. In the same way as soon as he has fully taken in and believed for himself that Jesus has borne our sicknesses, he does not fear to say: I no longer need bear my sickness; Jesus in bearing sin bore also sickness which is its consequence; for both He has made propitiation, and He delivers me from both" (p. 55).

This emphasis, though resisted and denounced by some evangelical leaders, began to exercise a fairly wide influence, especially among "deeper life" Christians, the so-called "holiness" groups, and later the Pentecostal groups. But not exclusively among these. Some within the historic denominations ardently accepted this teaching. A church of England rector, Rev. Howard Cobb, who was Warden of a Home of Divine Healing, wrote a book in 1933 which has had many reprints. Mr. Cobb takes a strong stand for healing through the atonement of Christ. He claims that "Christ bore our sickness in the same way as He bore our sins . . . He bore them as our substitute. . . . The bearing of our sicknesses is clearly shown to be a part of the work of the atonement" (*Christ Healing*, pp. 20, 21).

CRITICISM OF THE DOCTRINE

In spite of the fact that Dr. A. J. Gordon, Dr. A. B. Simpson, and Dr. Andrew Murray, as well as other pioneers of the concept of healing through the atonement, were well-known and revered conservative evangelicals, other conservative evangelicals strongly attacked their teaching. We have already quoted at some length from the well-known book by Professor Warfield. Another well-known evangelical who came out strongly, especially against healing in the atonement, was Dr. Rowland V. Bingham, General Director of the Sudan Interior Mission and editor of *The Evangelical Christian*. In a book published in 1921 and several times reprinted, *The Bible and the Body*, he gave special attention to the views of Drs. A. J. Gordon and A. B. Simpson.

In answering the question, *Is Healing in the Atonement?* Dr. Bingham deals particularly with Matthew 8:17: "Himself took our infirmities, and bare our sicknesses," which he calls the "*Magna Charta of the whole theory*" of Drs. Gordon and Simpson. He draws out two vital facts for the interpretation of this verse: (1) The taking of "our infirmities" and the bearing of "our sicknesses" was done, not at Calvary but at Capernaum, not at Christ's death, but during His life. He says "There is just the difference between bearing our sicknesses and bearing our sins that there is between Capernaum and Calvary. Christ bore the sicknesses and sufferings of mankind in His life, but our sins He bore in His death" (p. 55). (2) Matthew does not use the Greek equivalents for the Hebrew verbs of Isaiah 53:4, but substitutes verbs with a different meaning. Dr. Bingham writes, "Matthew deliberately drops the substitutionary word for *'bear'* which Isaiah uses in the verse quoted by him, and uses another word for *'bear'* which is never associated with propitiation or atonement. The word used by Matthew (Ebastasen), although quite common in the New Testament is never linked with atonement, but is employed to express sympathetic bearing, as, for example, when it occurs in Galatians 6:2, 'Bear ye one another's burdens,' or as in Romans 15:1, 'We that are strong ought to bear the infirmities of the weak.' "

Dr. Bingham also claims that Isaiah 53:4, 5, "By his stripes we are healed," refers to the healing of the soul from sin, not the healing of the body from sickness. He says, "It is a significant fact, too, that in every case where Isaiah uses the word 'health' or 'healing' in the prophecies of his book, he has spiritual and not physical health or healing in mind" (p. 57).

A later evangelical writer who dissented from the teaching of Doctors Gordon and Simpson was Dr. Henry W. Frost from whose book *Miraculous Healing* we have already quoted in another connection. Dr. Frost roundly declares that the argument for Christ's substitutionary bearing of human sickness "seems to be unscriptural." He says of Matthew 8:7, "There was no vicarious element in the Galilee healings, Christ not having yet suffered on the Cross, and hence universality cannot be founded upon them or deduced from them" (p. 58).

He explains Matthew's quotation of Isaiah 53:4 as follows: "It appears, therefore, that Isaiah 53:4, 5 was written with a double prophetic outlook: first to an atonement for sin, of which Peter speaks (1 Peter 2:24); and second, to the healing of disease, before and apart from the atonement, of which Matthew speaks (Matthew 8:17), this last, undoubtedly, as an evidence and proof of Christ's Messianic claim. This double significance, if a rightful interpretation is to be reached, must be kept in view, and the two must be held separate and must not be confused. In other words, Matthew 8:17 does not refer to the atoning work of Christ, and universal healing cannot be founded upon it. It refers to a temporary content connected with the earthly ministry of our Lord, which being 'fulfilled' was not to be renewed" (p. 59).

Thus among evangelicals in general two diametrically opposed views have developed during the last century concerning divine healing, and especially concerning healing through the Atonement. There are those reverent students of the Bible who fervently believe that Christ bore our sicknesses substitutionally even as He bore our sins. There are others, equally reverent in their attitude to Scripture, who strongly deny such a doctrine.

In view of this, it is not surprising that between the two schools there grew up a more careful attitude, and a more cautious expression, of the subject. One of these was Dr. R. A. Torrey. I came across an important quotation from Dr. Torrey in a book, *Divine Healing*, by Percy G. Parker, a

British Bible teacher, though I have been unable to verify the source of the statement. The statement is as follows: "The question arises, When do we get what Jesus Christ secures for us by His atoning sacrifice? We get the *firstfruits* of the atoning work of Christ, the *firstfruits* of salvation, in the life that now is, but we get the *full fruits* only when Jesus Christ comes again. The atoning death of Jesus Christ secured for us not only physical healing, but the resurrection and perfecting and glorifying of our bodies. No, we do not get the full measure of what Christ secured for us by His atoning death in the present life, but at His coming again. But while we do not get the full benefits for the body secured for us by the atoning death of Christ in the life that now is, but when Jesus comes, nevertheless just as one gets the firstfruits of His spiritual salvation in the life that now is, so we get the firstfruits of our physical salvation in the life that now is. We do get in many, many cases, physical healing through the atoning death of Jesus Christ, even in the life that now is" (quoted in *Divine Healing*, by Percy G. Parker, pp. 31, 32).

The interesting thing about this statement is its entire silence on the three controversial passages, Isaiah 53:4, 5; Matthew 8:17; and 1 Peter 2:24. Torrey looks upon physical healing as one of the *indirect* fruits of our Lord's death upon the Cross, rather than a matter of substitution. In this connection, Dr. Bingham quotes an interesting passage from Dr. A. J. Gordon: "In the atonement of Christ there *seems* to be a foundation laid for faith in bodily healing. *Seems*, we say, for the passage to which we refer is so profound and unsearchable in its meaning that one would be very careful not to speak dogmatically in regard to it —the passage *seems to teach* that Christ endured vicariously our diseases as well as our sins" (*The Bible and the Body*, by R. V. Bingham, p. 58). Bingham comments that A. J. Gordon "showed a hesitancy in this passage which is very foreign to his usual authoritative manner" (p. 58).

The Various Opinions of Pentecostals

It is, perhaps, through the work of Pentecostal evangelists and pastors that divine healing has received its most sensational emphasis. Not only has healing through the atonement of Christ been taught and believed within the sheltered quietness of Pentecostal churches and assemblies, but has been blazoned abroad through every form of mass media, radio, television and popular magazines and books, as well as proclaimed to multitudes in the largest auditoriums throughout the world.

However, when we come to examine the views of responsible Pentecostal leaders on the issue of healing in the atonement, we do not find that they all speak with a common sound. In fact, we find the same three points of view among them as we have noticed among Evangelicals in general.

There are those who fearlessly, even vociferously, proclaim that physical healing is in the atonement in exactly the same way as salvation from the penalty of sin. Just as Jesus died for our sins, so He died for our sicknesses. He suffered on the Cross as our substitute, both for sin and for sickness, and it is absolutely unnecessary for a Christian believer to bear either sin or sickness any longer. Sometimes, in the public crusades, language beyond all caution and restraint has been used to impress the masses with this message, and equally strong statements have been put into print. I quote one example only: "Jesus bore our infirmities, our diseases and our sicknesses and what He bore we do not need to bear. What He took upon Himself we do not need to suffer—Satan cannot legally lay on us what God laid on Jesus. He became sick with our diseases that we might be healed. He knew no sickness until He became sick for us. The object of His sinbearing was to make righteous all those who would believe on Him as their Sin-bearer. The object of His disease-bearing was to make well all those who would believe in Him as their Disease-bearer. . . . He took our diseases and so made us well; He took our infirmities, and so made us strong; and He now trades us success for our failures" (E. W. Kenyon, quoted in *Healing the Sick,* by T. L. Osborn, p. 37).

This, of course, is in complete harmony with the teaching of Gordon, Simpson and Murray.

At the very opposite extreme from this is a belief among some Pentecostals that there is absolutely no basis in the Bible for the idea that Jesus bore sickness substitutionally on the Cross. This point of view was advocated at a meeting of the Victoria Institute in London, England, in February, 1956, by a leading Pentecostal Bible teacher, L. F. W. Woodford. The lecture was subsequently published under the title: *Divine Healing and the Atonements: A Re-statement.*

Mr. Woodford carefully scrutinized the Biblical basis for the healing in the atonement view, and reached the conclusion that there is no foundation in the three passages in Isaiah, Matthew and 2 Peter for the doctrine.

In regard to Isaiah 53:4-5, Mr. Woodford argues that it has no connection with atonement for bodily sickness. Though Isaiah uses language which speaks of sickness and healing, this, he argues, is to be understood *metaphorically of sin.* He says, "The whole section of (Isaiah 53) is of one piece throughout. Its essential and repeated burden relates to sin, transgression and iniquity. . . . Thus Isaiah 53:4, in full keeping with the whole of this section of Scripture, declares the substitutionary work of the Suffering Servant of Jehovah for sin, set forth in terms of the stricken, smitten and afflicted leper" (p. 58).

On Matthew 8:17, Mr. Woodford says, "Matthew makes very significant changes in key words. His quotation entirely avoids any rendering into Greek of the substitutionary value of the Hebrew words used by Isaiah (bear—carry). Nor does he use the Greek verb 'to bear' (phero) used by the Septuagint in Isaiah 53:4. The latter verb is used in Scripture in a substitutionary sense (e.g. 1 Peter 2:24 and Hebrews 9:28). . . . But in place of phero, Matthew uses the verb bastadzo for 'bear,' which verb is *never* used in the New Testament in a substitutionary sense. This change of word is certainly arresting and is in keeping with the assertion that there is no thought of substitutionary sacrifice for sickness in the mind of Matthew in this Scripture. His quotation was related to the life-ministry of the Messiah, not to His sacrificial death, and his rendering of Isaiah was adapted accordingly and to definite purpose" (p. 59).

On 1 Peter 2:24 Mr. Woodford takes the view that there is in the whole context, no reference to bodily healing, but only to "the healing of the soul through the remission of sins" (p. 60). He also shows that in the statement of Peter, "By whose stripes ye were healed" (1 Peter 2:24), quoted from Isaiah 53:5, there is absolutely no thought of the scourging of the Lord Jesus after His trial. He says, "It is very significant that Peter did not use any of the three words employed in the New Testament for beating, flogging or scourging (Luke 18:33; John 19:2; Matthew 27:26; Mark 15:5). . . . If he had desired to refer to the scourging of the Lord Jesus he would surely have used one of these appropriate words, but in fact he did not do so. The statement he made did not refer to the scourging of the Lord Jesus, but to the stroke of death laid upon Him by God on our behalf" (pp. 55-61).

It should be stressed that Mr. Woodford does not surrender the doctrine of supernatural divine healing. He is critically examining the concept of healing in the atonement, and comes to the conclusion that bodily healing comes to us through the risen life of Christ rather than through His substitutionary sacrifice. As he puts it, "The virtue of the atoning blood of Christ has released the power of His risen life for the physical need of man" (p. 61).

As with all other evangelicals, there are those Pentecostals who are more cautious in their statements, though they clearly believe in healing through the atonement. One of these was the late George Jeffreys, one of the most outstanding British evangelists between the two world wars. He filled the largest auditoriums in Britain and many parts of Europe, leading large numbers to Christ for salvation and healing, yet, both in his teaching and practice, he was singularly free from extravagances and fanaticism.

On the question of healing through the atonement, he clearly stated his belief in this doctrine but carefully avoided any suggestion of substitutionary language. In his book, *Healing Rays,* first published in 1932, he makes the following observations: "If sin, the cause of all the evil effects from which the creation is suffering, could only be put away by the atoning work of Christ upon the Cross, then it is only reasonable to conclude that provision has been made in that atoning and redeeming work for the putting away of its effects" (p. 23). Later in his book he poses the question: "Is healing for the body in the atonement?" He answers, "The atoning and redeeming work of Christ on the Cross is the sovereign remedy for all the evil effects of the first Adam's disobedience, which include sickness and disease. Bodily healing is one of the present day benefits of that atoning and redeeming work" (p. 154). The interesting thing here is that this convinced and much-used believer in healing through the atonement of Christ does not use the language of substitution or quote any of the three fundamental texts which A. J. Gordon, A. B. Simpson and others laid such heavy stress on as a scriptural basis for the doctrine. Indeed, all through his book he seems studiously to avoid these texts. He merely says that if the atonement does away with sin, then, logically, it does away with the effects of sin, among which are sickness and disease.

Another British Pentecostal Bible teacher who held a mediating position on this matter was Percy G. Parker. In his book, *Divine Healing,* he teaches that though bodily healing is in the atonement, it is not *directly* in the atonement, but only *indirectly* so. He says, "Isaiah 53:4 was fulfilled in Christ's *life*. It is true that it was in view of the death on the Cross that He was able to remove sickness in His life. But to say that He removed sickness in virtue of His Cross is a different thing from saying that He removed sickness by being made sick on the Cross, or by bearing stripes for physical sickness" (p. 31). Again, he says, "Christ's death was the ransom price. Now, being saved by His death, the Lord Jesus heals by His life. It is the life of Christ which brings health. It was the death of Christ which brought atonement. The poured-out blood saves the sinner from guilt. The poured-in resurrection life saves the saint from disease. Forgiveness through the atonement is God's gift to the sinner. Health through the life of the resurrected Christ is God's gift to the saint" (p. 33). Later in his book he deals with a questioner who challenged his conception of healing *indirectly* through the atonement. He says, "There was no need for Christ on Calvary to suffer because of sickness, for sickness is simply the fruit of the penalty of sin. The penalty of sin was separation from God which penalty Christ suffered. If sickness were part of the penalty of sin, not part of the fruits of the penalty of sin, then every one, the moment Christ was accepted, would not only flash into immediate communion with God but also into immediate perfect health" (p. 65).

From our examination of these various opinions, both among Pentecostals and among evangelicals in general, it is evident that more serious thinking is needed on the subject of healing in the atonement. It would be a most excellent thing if a study commission could be set up, made up of responsible Bible scholars from the leading Pentecostal denominations, to examine this subject thoroughly and honestly. Perhaps, they could produce a clearly-worded statement which would preserve the basic truth of "healing as in the atonement," yet be free from the ambiguities and anomalies which have beclouded it.

In my closing remarks I should like to make a few suggestions in regard to this. These will be my personal opinions only. They are only tentative solutions. I offer them for your consideration.

A Tentative Solution

I shall begin by saying that the interpretation of the three cardinal texts (Isaiah 53:4, 5; Matthew 8:17; and 1 Peter 2:24), *which is advocated by Drs. Bingham and Frost and Mr. Woodford, is, in my opinion, a sound one.* As we have seen, these are the usual passages put forward to support the teaching that Jesus bore sickness vicariously and substitutionally, just as He bore our sins. But a careful exegesis of these texts makes it clear that they do not teach what some have claimed. Isaiah 53:4, 5 and 1 Peter 2:24 teach that Jesus bore our sins on the Cross, but they do not specifically refer to bodily sickness. Matthew quotes the passage from Isaiah and applies it to bodily sickness, but he makes a change in the vital verbs, and applies it to the ministry of Jesus during His life, not at His death.

Having said that, I want, however, to state that I cannot accept the view of Professor Warfield and Dr. Henry Frost that the healing miracles of Jesus and the Apostles were mainly, if not indeed, only, demonstrations of His Messiahship and Deity. There is truth in that view, but it is not the whole truth. There is another element present in the healing miracles of Jesus. *This is the element of compassion.* I referred to this in our first lecture when quoting Dr. Plumtre. Let me here quote the Gospel statements: Matthew 9:36, "But when he saw the multitudes he was moved with compassion on them," (also Mark 6:34); Matthew 14:14, "And Jesus went forth, and saw a great multitude and was moved with compassion towards them, and he healed their sick." Matthew 15:32, "And Jesus called his disciples unto him, and said, I have compassion on the multitude" (this is in connection with the hungry multitude; see also Mark 8:2); Matthew 20:34, "And Jesus had compassion on them (two blind men), and touched their eyes." Luke 7:13, "And when the Lord saw her (the widow of Nain), he had compassion on her" and raised her son from the dead. Mark 1:41, "And Jesus, moved with compassion put forth his hand, and touched him, and saith unto him, I will, be thou clean." Mark 5:19, Jesus said to the demon-possessed man whom He had delivered, "Go home to thy friends, and tell them how great things the Lord hath done for thee, and hath had compassion on thee."

This emphasis on compassion as a motivating force of our Lord's ministry of miraculous healing is extremely important and very impressive. The Greek word translated *compassion* in all these quotations adds to the impressiveness. The verb comes from the noun *splagchna,* which is translated by the word *bowels* in the King James Version. Out of nine occurrences it is used literally only once in the New Testament (Acts 1:18). In the other eight occurrences it has a metaphorical significance (2 Corinthians 6:12; Philippians 1:8; 2:1; Colossians 3:12; Philemon 7, 12, 20; and 1 John 3:17). I quote only one, Philippians 1:8: "For God is my record, how greatly I long after you all in the bowels (*splagchna*) of Jesus Christ." J. B. Lightfoot says, "The *splagchna* are properly the nobler *viscera,* the heart, lungs, liver, etc., as distinguished from the *entera,* the lower *viscera,* the intestines." From this it is evident that the word *bowels* of the K.J.V. is not a good rendering into English. Modern versions use various words, such as *heart* and *affections.* For example, in Philippians 1:8, the R.S.V. has, "I yearn for you all with the affection of Jesus Christ." Phillips has, "I long with the deepest Christian love and affection." The New English Bible has "I long for you all with the deep yearning of Christ Jesus Himself." Lightfoot comments, with Paul especially in mind, "The believer has no yearnings apart from his Lord; his pulse beats with the pulse of Christ; his heart throbs with the heart of Christ."

Dr. William Barclay, in his book, *New Testament Words,* has an excellent study on this word.

He says, "In classical Greek the *splagchna* means the inner parts of man, which are the seat of the deepest emotions. It is from that idea that the verb *splagchnidzesthai* was formed in later Greek. It means to *be moved with compassion,* and from its very derivation, can be seen that it describes no ordinary pity or compassion, but an emotion which moves a man to the very depths of his being. It is the strongest word in Greek for the feeling of compassion" (p. 276).

I have gone into this at some length in order that we might see what the Gospel narratives mean when they tell us so often that Jesus on His healing ministry was *"moved with compassion."* He was motivated by deep and powerful feelings of pity and compassion. And this is only another way of saying that He was motivated by love, for compassion is an expression of real love.

Any interpretation of the healing miracles of Christ must give a large place to this powerful impulse of compassionate love. We cannot say that He was merely validating His deity or that the Father was merely validating His Son's deity. That factor is certainly present. But equally is the factor of compassionate love present. Indeed it is essentially present for only a ministry which is motivated by compassionate love can be a true and genuine Messianic and divine ministry.

The important point I wish to make is that once we have recognized this fundamental element of compassionate love in Christ's ministry of healing we have gone a long way towards acknowledging that the healing of men's bodies is indeed, deeply involved in the motives and reasons for His death upon the Cross. The same powerful motive of compassionate love which moved Him to involve Himself during His life in the sicknesses and sufferings of men and women for their deliverance, also moved Him to involve Himself in the sin and guilt of mankind in His death upon Calvary. There were not two different motivations, one for sickness and another for sin, but the one great divine motivation of compassionate love, behind the sickness-bearing ministry of His life and the sin-bearing mission of His death. The Father and the Son and, indeed the Holy Spirit too, are deeply and profoundly concerned and moved (a moving which is powerfully portrayed by the word *splagchna*) by the terrible sins and sufferings and sicknesses of mankind—and these things are not separate and utterly distinct items but are all of a piece and interrelated. It was this compassionate love which purposed the great plan of redemption. It was this compassionate love which brought about the Incarnation of the Son of God. It was this compassionate love which moved Jesus to involve Himself in the sicknesses and sorrows of men and women in His brief three-year ministry. It was this compassionate love which led Him finally to the Cross to "die, the just for the unjust, that he might bring us to God." And because He is "the same yesterday and today and for ever," this compassionate love still moves and motivates Him from His throne in heaven, and by His Spirit and through His Church on earth.

This may not be what the word *atonement* technically means. Nevertheless, it is not very far removed from it. The basic motivations are the same. This basic motivation is in Isaiah 53 as well as in Matthew 8:17. In this sense, healing is certainly in the atonement.

I wish to make a third observation on this important theme: I believe that physical healing is related to the death of Christ in a more specific way. It is involved in the New Testament concept of "redemption." Various terms are used in the

New Testament to describe the purpose and effects of the death of Christ. Dr. W. H. Griffith Thomas lists the six leading words in his book, *The Principles of Theology* (pp. 52-53). They are: 1. Sacrifice; 2. Offering; 3. Ransom; 4. Redemption; 5. Propitiation; 6. Reconciliation.

It will be noted that the word *atonement* does not occur here. Actually, the word occurs only once in the King James Version of the New Testament, in Romans 5:11, "By whom we have now received the atonement." There the original would be better rendered "reconciliation" as in the R.S.V. *atonement* is not a Greek word, but pure English. As used in theology, it seems to be a general term which comprehends and sums up all the six New Testament terms which Dr. Thomas describes. It describes the death of Christ upon the cross as the divine means of our salvation. It includes all the ideas of sacrifice, offering, ransom, redemption, propitiation, and reconciliation.

The words to which I wish to draw attention are *ransom* and *redemption*. The underlying Greek words are *lutron* and *apolutrosis*. These are nouns. A related verb is *lutroo*. There is a close connection between these words. Dr. G. Abbott Smith in his *Manual Greek Lexicon* defines them as follows: "*Lutron* means *ransom; lutroo* means to release on *payment of a ransom*, or to *release by paying ransom*, or redeem, or deliver. *Apolutrosis* means *to release on payment of a ransom*, or *release effected by payment of ransom, redemption, deliverance*." Cremer enriches this somewhat by saying that *lutron* "almost always means the price paid *for the liberation of those in bondage*," and that *lutroo* "denotes that aspect of the Saviour's work wherein He appears as the Redeemer of mankind from bondage." Dr. Thomas also says concerning the word *apolutrosis*: "The *thought appears to be the removal of bondage and thraldom."*

Let us now look briefly at the New Testament uses of these words: *lutron* occurs twice, both times in connection with the death of Christ, Matthew 20:28 and Mark 10:45, the same statement in each case, "to give his life a ransom for many"; the clear meaning is that the death of Christ was the price paid for the deliverance of mankind. The statement, however, does not tell us what the deliverance is from. The verb *lutroo* occurs three times, two of which have to do with the death of Christ, and they tell us what the deliverance is from. The verses are: Titus 2:14 and 1 Peter 1:18. Titus reads "Christ gave himself for us, that he might redeem us from all iniquity." This is an interesting statement because it tells us that the purpose of the death of Christ was our deliverance from the *power* of sin; not merely its penalty, but its power. W. E. Vine comments that the Greek word *anomia*, translated "iniquity," means "the bondage of self-will which rejects the will of God." Thus the statement in Titus 2:14 is a clear "sanctification text." Peter reads, "Forasmuch as ye know that ye were redeemed not with corruptible things from your vain manner of life—but with the precious blood of Christ." Dr. James Denney, in his book, *The Death of Christ*, has an interesting statement on the clause "redeemed . . . from your *vain manner of life.*" The Greek word translated "vain" is *mataia*, and Denney says, "Life before the death of Christ is *mataia*, that is, it is futile, it is a groping or fumbling after something it can never find; it gets into no effective contact with reality; it has no abiding fruit. From this subjection to vanity it is redeemed by the blood of Christ. When the power of Christ's Passion enters into any life it is not futile any more; there is no longer the need or the inclination to say 'all is vanity.' Nothing can be more real or satisfying than the life to which we are introduced by the death of Christ; it is a life in which we can have fruit, much fruit, and fruit that abides. Hence the introduction to it, as the word *elutrothete* (redeemed) suggests, is a kind of emancipation" (pp. 92, 93). Though neither *lutron* nor *lutroo* are specifically applied to physical healing in the New Testament, I hope to show that the emancipation achieved for mankind through the ransoming power of Christ's shed blood includes healing for the body.

Now we must look at the noun *apolutrosis*. It occurs ten times in the New Testament, all but one of these occurrences being in the Pauline epistles and Hebrews. Nine of these references refer to redemption through Christ. We will now examine each of these, as there are different shades of meaning among them.

On two occasions the word is used *in a general sense* of our total salvation: Romans 3:24, "Being justified freely by his grace through the redemption that is in Christ Jesus"; Hebrews 9:15, "And for this cause he is the mediator of the New Testament, that by means of death, for the redemption of the transgressions that were under the first testament, they who are called might receive the promise of eternal inheritance."

On two occasions *apolutrosis* is given *the limited meaning of forgiveness of sins*: Ephesians 1:7, "In whom we have redemption through his blood, even the forgiveness of sins, according to the riches of his grace"; Colossians 1:14, "In whom we have redemption through his blood, even the forgiveness of sins."

Once the word is used to describe *deliverance from the disasters which are to come upon the world at the consummation of the Christian age*: Luke 21:28, "And when ye see these things begin to come to pass, then look up and lift up your heads for your redemption draweth nigh."

The remaining four occurrences have to do with our *final redemption at the Second Coming of Christ, which is clearly revealed in Scriptures to include the glorification of the body*. Romans 8:23 is the plainest and clearest statement, "And not only they, but ourselves also, who have the firstfruits of the Spirit, even we ourselves groan within ourselves, waiting for the adoption, that is, the redemption of our body." There are two statements in the Epistle to the Ephesians 1:14, where Paul describes the Holy Spirit as "the earnest of our inheritance, until the redemption of the purchased possession, unto the praise of his glory"; and 4:30, in similar language, "And grieve not the Spirit of God, by whom ye are sealed unto the day of redemption." The last of these four passages is in 1 Corinthians 1:30, "But of him are ye in Christ Jesus, who of God is made unto us wisdom and righteousness, and sanctification, and redemption." Expositor thinks that the word "redemption" here "covers the entire work of salvation," but Wesley declares that it has to do with the consummation of our salvation at the Second Coming of Christ. He takes "righteousness" to mean *justification;* "sanctification" to refer to the experience of inward *holiness;* and "redemption" to be *"complete deliverance* from all evil, and eternal bliss, both of soul and body." The use of these three words, "righteousness, and sanctification, and redemption" is very striking and suggestive here, and inclines us to believe that Paul is referring to the "redemption of the body" in the third of these terms.

I have gone at some length into the use of

these New Testament words for "redeem" and "redemption" because I am convinced that if we are to find an adequate basis for the concept of "divine healing in the atonement" it must be within the thought of "redemption through the blood of Christ."

The Son of God offered Himself upon the Cross in order that He might provide "redemption" for lost humanity. The New Testament unmistakably teaches that the shedding of His blood was the ransom price which He paid. From our study of the various references it is clear that by His death upon the Cross Christ purchased the emancipation, the deliverance, the freedom, of all who put their trust in Him, from all of the tragic consequences of Adam's fall. The New Testament itself shows that these tragic consequences are threefold: (a) man is guilty and condemned and under the wrath of God; (b) he is held in the grip of the spirit of lawlessness or self-will, variously called "the old man," "the carnal mind," "the flesh," etc. (c) His body is subject to death and corruption, including all of the sicknesses, diseases, and infirmities which are part of the process of dying. The "redemption" which Christ purchased for us is an all-inclusive deliverance from all of those consequences. It includes *justification* and the "forgiveness of sins"; it includes *sanctification*, or "redemption from all iniquity or lawlessness"; and it includes *glorification*, or the "redemption of the body."

Thus, the New Testament unmistakably teaches that the salvation which God has made available to a lost and fallen world through His grace in Christ, is a salvation for spirit, soul, and body. The body is not a useless and despised thing to be ultimately discarded and destroyed. It is included in the redemption which Christ purchased for us on the Cross. Jesus died to provide both a spiritual and a physical salvation.

Now there are two features about this physical salvation which Paul presents to us in Romans 8. First, it is a future hope rather than a present experience. It is "a glory to be revealed in us" (verse 18). We are still "waiting" for it (verse 23). We are hoping and waiting with "patience" for it (vv. 20-25). These are statements which provide a sufficient refutation to those who vainly imagine that if they have faith they will never experience any pain or weakness or physical limitation whatsoever. The "redemption of the body" is a future blessing rather than a present experience.

But this is only one side of the antinomy. Second, Paul says that Christian believers have "the firstfruits of the Spirit" (v. 23). This is an illuminating metaphor. The "firstfruits" are the very early fruits which precede the great ingathering of harvest. The ministry of the Holy Spirit in the life of the Christian is exactly that. It is a prelude and foretaste of the life and glory which will come to us at the return of Christ. There is another term in the Pauline letters which is similar in meaning: the term *"earnest"* in 2 Corinthians 5:5 and Ephesians 1:13, 14—*"the earnest of the Spirit."* The word means a first installment, a deposit, a pledge. The Greek word, *arrabon,* is used by modern Greeks for an engagement ring.

There are many ways in which we may experience this firstfruits and this foretaste. Inasmuch as Paul in Romans 8 has "the redemption of the body" specifically in mind, I think that one of these ways is in regard to the body. Not only may we enjoy the blessedness of justification and of sanctification, but also we may taste in advance the firstfruits of "the redemption of the body." We may do so in two ways:

By a miracle of healing. When Jesus healed the sick He was giving to them a "firstfruits" of "the redemption of the body" which He was to make possible by His death. So with the healing ministry of the Apostles and their helpers: it was an "earnest" and "firstfruits" of final physical salvation. Every miracle of healing that the risen Lord has given through His Church on earth is a deposit and pledge of that "redemption of the body" which He has purchased by His death upon the Cross. Many years ago, David Baron pointed this out in his exposition of Isaiah 53:

"The miracles of healing not only served to certify Christ as Redeemer, and as 'signs' of the spiritual healing which He came to bring, but were, so to say, pledges also of the ultimate full deliverance of the redeemed; not only from sin, but from every evil consequence of it in body as well as in soul. Hence our full salvation includes not only the perfecting of our spirits, but the 'fashioning anew of the body of our humiliation that it may be conformed to the body of His glory'" (*The Servant of Jehovah,* p. 86, by David Baron).

But not every person who turns to Christ for physical healing is healed; not even every Christian. What then? Does this mean that God gives nothing but a blunt refusal to our prayers?

I believe that the case of Paul's "thorn in the flesh" sheds light on this matter (see 2 Corinthians 12:7-10). There are divergent opinions concerning this "infirmity." My own considered view is that it was a physical weakness or disease of some kind. Paul tells us that the Lord did not deliver him of it. But this is not the whole story. God did not give a blunt refusal. He gave him something equally as wonderful, equally as positive, equally as supernatural, as a miracle of healing. He said, "My grace is sufficient for thee: for my strength is made perfect in weakness" (v. 9). And Paul replied, "Most gladly, therefore, will I rather glory in my infirmities, that the power of Christ may rest upon me" (v. 9). Though he was not healed, he was given a grace that enabled him to live triumphantly with his weakness. He was brought from under its domination. He was delivered from resentment, and from self-pity, from fear, and all of the psychological disorders that frequently accompany chronic sickness and infirmity. He was enabled to transcend his physical limitations. And thus God was glorified equally by this manifestation of transcending grace, as by a physical miracle.

In either case, it is a miracle. In either case, it is a deliverance, a redemption. In either case, it is one of the wonderful blessings that come to us through the atonement.

I should like to conclude with a quotation from a helpful little booklet by Douglas Webster, called *The Healing Christ*. Mr. Webster does not have physical healing exclusively in mind in his short studies, but he certainly includes it. In this quotation he really has total salvation in mind, the healing of body, soul and spirit. He says, "In Paul's thought Christ is like Adam, He is the Second Adam, the Head of a race, an incorporating personality. As all men were affected by Adam's sin, as all died in Adam, became bound as he did, so—claims Paul—all have potentially died in Christ, and all are affected by what He has done, by this great healing deed. In other words, the potentiality of Adam, of all ordinary humanity, has been neutralized and exceeded by the potentiality of Christ. The glorious deed that Jesus did is big enough to undo all the consequences of the terrible deed that Adam did and the ghastly things that the sons of Adam, all of us, still go on doing. So Christ's death, His Cross, is the great healing deed. In His wounds we find our **safety**, in His stripes our **cure**, in His pain our **peace**, in His **Cross** our victory. Here and nowhere else is healing for all the diseases of our souls'' (pp. **44, 45**).

Chapter IV

APPROPRIATING AND COMMUNICATING THE TRUTH

I have been acutely conscious all through these discourses that we are grappling with problems which are by no means easy. Indeed, throughout my ministry, covering over thirty-five years, I have increasingly felt both the greatness and the complexity of the matters which we have discussed. But as I approach the final lecture I have a particular feeling of inadequacy and weakness. The reason for this is, no doubt, that of all problems relative to the truth of divine healing none is greater than the problem of appropriation and communication.

Of course, appropriation and communication are vital for all knowledge and all concepts. Unless ideas can be communicated to other minds and appropriated by other minds there is bound to be utter frustration. In regard to Christian truth, however, the frustration is much greater. For Christian truth is no mere theory, neither is it a merely temporal or material thing. It is supremely spiritual, eternal and practical. We are not merely seeking to appropriate and communicate theories, ideas, concepts, but those experiential realities which the concepts represent. Not only do we endeavour to lead people to apprehend the great *ideas* of Justification, Regeneration, Sanctification the Pentecostal Baptism, and Divine Healing, but to enter by faith into the realities which those wonderful words describe.

In this final lecture we are to try to face up to this, the greatest of all problems, surely, in regard to the truth of divine healing: how to appropriate and communicate, not the concept, merely, of divine healing, *but divine healing itself.* Two things are involved here: (1) *Communication has to do with others.* What a challenging problem this is! All around us are multitudes of people in great distress, suffering from all manner of diseases. Within the Christian churches, even within the churches which hold to the truth of divine healing, there are large numbers of sick people. How to communicate divine healing to these people is surely a matter of great and grave concern. (2) But there is also the question of *appropriation.* This has to do with *myself;* and by *myself* I mean the pastors and evangelists and church leaders who hold to the truth of divine healing. Many of these are themselves sick, some battling with chronic and crippling diseases, others being suddenly incapacitated by acute, killing diseases. Ofttimes, sickness comes to their families. Sometimes, the very people who seem to be greatly used in communicating divine healing to others, are themselves victims of sickness and infirmity. How can they, how can we, how can I, appropriate personally the wonderful blessing of divine healing?

I shall not attempt to differentiate between these two classes, between appropriation and communication, but I will attempt to describe some principles which are of paramount importance for both.

WE SHOULD TRY TO CREATE HOPE IN THE MIND OF THE SUFFERER

The late Dr. A. T. Scofield, a Christian Harley Street specialist, wrote: "In disease, functional or organic, the therapeutic value of faith and hope, though not in our textbook, is often enough to turn the scale in favour of recovery" (*Nervousness*, p. 61). This well-known fact was illustrated a few years ago at the annual meetings in London, of the *Mission to Lepers*, by Dr. A. W. Davies. He declared that, "The creation of hope was the most important factor in the cure of leprosy." He described an educated man in a leper home, and said of him, "The Christian atmosphere of that place had arrested the patient's deterioration and had roused in him the hope and will to fight the disease."

In a much more profound way we can see this in the ministry of Christ. His wonderful personality and character and teaching and miracles created hope in multitudes of hopeless lives. People who had no hope of seeing again, or of hearing again, or of walking again, lepers who had no future but a weary waiting for a merciful death, suddenly became aware that a new factor had come into their hopeless lives. "The people that sat in darkness saw great light" (Matthew 4:16). And this new hope was the first step towards a miraculous recovery. To take but one example, the case of blind Bartimaeus (Mark 10:46-52). How long he had "sat by the wayside begging" we do not know. Most of his life probably. But suddenly there is a new factor in his situation. He had heard the sensational reports of Jesus of Nazareth and His mighty deeds. As he sits there on this glorious day he hears the jubilant excitement of a great crowd of people as they approach the city of Jericho, and learns that Jesus of Nazareth is with the crowd. Immediately hope catches fire in his breast and he begins to cry out, "Jesus, thou Son of David, have mercy on me." The bystanders cannot silence his cries of expectation and hope, and he cries out until the attention of Jesus is gained.

Now, of course, more than this was required for his healing. But the creation of hope and expectation was a primary factor leading to a miracle of deliverance.

It was so, also, with the Apostles. Because "they had been with Jesus" and Jesus was "with them," they were able to inspire disease-stricken people with the hope of recovery. Take, for example, the narrative in Acts 5 concerning the people of Jerusalem. Luke tells us that "they brought forth the sick into the streets, and laid them on beds and couches, that at the least the shadow of Peter passing by might overshadow some of them" (v. 15). This was not a "healing technique" of the Apostles. We are not told that the Apostles encouraged it, or even that healing came to them through Peter's shadow. It was the people themselves who acted like this. And why did they do it? Because hope of *healing had been created within them.*

Here we have the first stage in a healing ministry or a healing experience. We must create hope of healing in the mind and heart of the sufferer. And if we are true followers of the Lord Jesus and His Apostles, this is surely what we shall endeavour to do for men and women, whatever may be the kind or the depth of their need. However hopeless their situation may appear to be we have good news for them. We bring a new factor into their situation. We focus their attention upon God —the Living God of the Bible. God loves them. God cares for them. God is concerned about them. God has the answer to their problem and their need. This factor alone, if truly communicated and appropriated, is enough to "turn the scale in favour of recovery."

But this is not all. We can draw the attention of the sufferer to the great promises of divine healing in the Bible: "I am the Lord that healeth thee" (Exodus 15:26); "Bless the Lord, O my soul, and forget not all his benefits: who forgiveth all thine iniquities; who healeth all thy diseases" (Psalm 103:2-3). "They shall lay hands on the sick, and they shall recover" (Mark 16:18); "Is any sick among you? let him call for the elders of the church, and let them pray over him, anointing him with oil in the name of the Lord; and the prayer of faith shall save the sick, and the Lord shall raise him up" (James 5:14-15).

We can go on to mention the wonderful healing compassion of Christ, to which we have previously referred. We can point out that if Jesus Christ is, as Scripture declares, "the same yesterday and today, and for ever" (Hebrews 13:8), then He is still being moved with compassion towards sick and diseased humanity. He is still, as Hebrews 4:15 says, "touched with the feeling of our infirmities."

We can go further still. We can show to the sick and suffering that there is, indeed, healing in the atonement. We can show that on the Cross He died as our Ransom and our Redeemer. He paid the price to redeem and save us from all the consequences which have come upon the human race through Adam's sin and fall. That redemption and salvation is a full redemption and salvation, for the whole man, spirit, soul, and body. While we cannot enter into all that it means until Jesus Christ comes back again, yet we can receive the firstfruits of it in forgiveness, and cleansing, and power, and healing, and strength for the body.

If we will faithfully preach and teach and talk about these great truths we will radiate hope to hopeless people on every level of need. And if we will habitually and believingly think about these truths ourselves, "meditating upon them day and night," we too, as pastors and evangelists and leaders, will counteract the unbelief and despair that sometimes gnaws at our own hearts.

Before leaving this matter of the creation of hope, I feel I ought to say something about the techniques of advertising. How far are we justified in using the techniques of advertising to create hope in the minds of sick people? We live in an age of mammoth publicity. Advertising is very big business, and there is no doubt that it is a key factor in commercial success. This has been recognized by church leaders, especially in connection with evangelism. It is easy also to see how the publicizing and advertising of divine healing have tremendous appeal to people who are hopelessly sick. How far are we justified in using this technique to create hope in the minds of sufferers?

In recent years advertising has gone into a new dimension—the dimension of *depth psychology advertising*. Instead of appealing to our reasoning faculty, as in the old days, the advertisers now appeal to our subconscious motives—our fears, hopes, desires, anxieties, etc. In a gripping book, *The Hidden Persuaders,* Vance Packard had made a searching investigation of this modern technique of advertising, especially in America. He writes, "Large scale efforts are being made, often with impressive success, to channel our unthinking habits, our purchasing decisions, and our thought processes by the use of insights gleaned from psychiatry and the social sciences. Typically, these efforts take place beneath our level of awareness; so that the appeals which move us are often, in a sense, 'hidden.' The use of mass psychoanalysis to guide campaigns of persuasion has become the basis of a multi-million-dollar industry. Professional persuaders have seized upon it in their groping for more effective ways to sell us their wares—whether products, ideas, attitudes, candidates, goals, or states of mind" (p. 11). He goes on to describe all of this in various kinds of industries and commercial enterprises, and also in professional polities, and even in religion. He says, in a foreword, that as a result of this modern technique of depth advertising, "Americans have become the most manipulated people outside the Iron Curtain." Those of us who live in Britain are seeing a rapid growth of the same thing.

In regard to religion, Mr. Packard says a significant thing: "Public-relations experts are advising churchmen how they can become more effective manipulators of their congregations" (p. 13). This is a matter of serious concern to Bible-believing Christians. The depth-psychology advertisers are not concerned about the *truth* of what they are telling the public over radio and television and cinema-screen, and in the newspaper ads. The one and only criterion is: *Is it successful? Will it sell? Will this advertisement condition the minds of people to buy this or that, to do this or that, to vote this or that, to believe this or that?*

True Christians cannot engage in this sort of manipulation, whether it be in regard to evangelism, soul winning, divine healing, or any thing else. We are governed by truth. Truthfulness is the deciding factor. When Martin Luther stood before the Diet of Worms he declared, "My conscience is bound by the Word of God." In a similar way, we are bound by the truth. The same truth that sets us free, also, and at the same time, *binds us*. In our preaching, teaching, propaganda, advertising, or in any thing else, we are limited by the truth. We can only operate within the frontiers of truth.

I believe that we ought to respect this principle in regard to divine healing. We earnestly wish to create hope in needy lives. But let us beware of exploiting their suffering in order to *sell* them something. We are not trying to *sell* them anything —we are trying, with the compassion of Christ, to help them. We can only do this by staying within the boundaries of truth. If we go beyond truth, whether in our preaching or teaching or reporting or advertising, we shall create, not a living hope, but illusions which will ultimately rebound in frustration and misery.

IN THE QUEST FOR HEALING THERE IS A PLACE FOR INTERCESSORY, IMPORTUNATE, BELIEVING PRAYER

If there is one thing that is quite clear it is that the gifts of God do not usually come to man automatically and unconditionally. There are recorded cases where men seemed to do nothing and God did everything, but usually the man himself has to cooperate with God in order to appropriate His blessings for himself, or to communicate them to others.

One of the primal ways of cooperation with God is by prayer. Prayer is a fundamental, divinely-ordained way of making contact with God, of entering into communion with God, of becoming both a receiver and a transmitter of His power. We may not know a great deal about the theological and psychological laws of prayer, but if we will give ourselves to the practice of prayer we will certainly prove for ourselves its vitality and power.

The Bible is far from silent about prayer in relation to divine healing. Let us take note of some of the truths it teaches.

By precept and by example the sufferer is encouraged to pray to God for deliverance. How frequently in the Bible we are urged to "call upon the Lord" in our time of need: "Call upon me in the day of trouble: I will deliver thee, and thou shalt glorify me" (Psalm 50:15). "Fools because of their transgressions and because of their iniquities, are afflicted. Their soul abhorreth all manner of meat; and they draw near unto the gates of death. Then they cry unto the Lord in their distresses. He sent his Word, and healed them, and delivered them from their destructions" (Psalm 107:17-20). "The sorrows of death compassed me, and the pains of death gat hold upon me: I found trouble and sorrow. Then called I upon the name of the Lord; O Lord, I beseech thee, deliver my soul" (Psalm 116:3-4).

In the Gospels there are some notable examples of desperately needy people praying to Christ for healing. There is the leper who came to Him, "beseeching him, and kneeling down to him, and saying unto him, If thou wilt, thou canst make me clean" (Mark 1:40). There was Jairus, a Synagogue ruler, who "fell at (Christ's) feet, and besought him greatly, saying, My little daughter lieth at the point of death: I pray thee, come and lay thy hands on her, that she may be healed; and she shall live" (Mark 5:22-23). There was the Centurion whose servant was dying, and who sent the elders of the Jews, "beseeching him that he would come and heal his servant" (Luke 7:2, 3). There was the father of the demon-possessed boy who met Jesus at the foot of the Transfiguration mountain, and "cried out, saying, Master, I beseech thee, look upon my son: for he is mine only child" (Luke 9:37-40). The Gospel narratives abound in such stories, and every one of them is an encouragement to us to pray to God for divine healing.

There are some fundamental statements about prayer in the Johannine account of Christ's teaching which offer the utmost encouragement to the sick and suffering: "Whatsoever ye shall ask in my name, that will I do, that the Father may be glorified in the Son. If ye shall ask any thing in my name, I will do it" (John 14:13, 14). "If ye abide in me, and my words abide in you, ye shall ask what ye will, and it shall be done unto you" (John 15:7). "Verily, verily, I say unto you, Whatsoever ye shall ask the Father in my name, he will give it you" (John 16:23). Can we doubt that the words *whatsoever, anything,* and *what ye will,* include our prayers for divine healing?

Christ did not teach that such petitions would immediately bring answers, like a slot-machine or a computer. He taught us to be persistent and patient in our praying. You will recall the two great parables in Luke's Gospel, the Parable of the Importunate Man who aroused his friend at midnight in order to borrow bread (11:5-8), and the Parable of the Importunate Widow who troubled the unjust judge with her cries for justice until he acted for her (18:1-7). The purpose of both these parables is plainly to show us that our prayers are not always answered at once, but only after patient and persevering intercession. As Jesus Himself said, "Go on asking, and it shall be given you; go on seeking, and ye shall find: go on knocking, and it shall be opened unto you" (Luke 11:9, Greek tense).

Another fundamental feature of Bible teaching on prayer which has special relation to divine healing is intercessory praying for others. The passages we have just quoted seem to apply mainly, if not exclusively, to praying for oneself. Sometimes, however, a sick person may be so depressed and weak as to be unable to engage in importunate praying. He becomes almost wholly dependent on the prayers of other Christians. And the New Testament urges him to solicit the prayers of other Christians. And the New Testament urges him to solicit the prayers of God's people. "Is any sick among you? let him call for the elders of the church; and let them pray over him, anointing him with oil in the name of the Lord: and the prayer of faith shall save the sick, and the Lord shall raise him up; and if he have committed sins, they shall be forgiven him. Confess your faults one to another, and pray one for another, that ye may be healed. The effectual fervent prayer of a righteous man availeth much" (James 5:14-16).

This is a tremendous statement concerning prayer. It is a whole guidebook on prayer concentrated in a few dynamic sentences. And, let us note, it especially concerns the subject of divine healing. James describes for us in letters of light and flame what kind of prayers they are which bring deliverance to the sick and afflicted, and what kind of people are able to pray those prayers.

Notice, that he has the ministers of the church in mind, for *"the elders of the church"* is a general New Testament term for pastors, evangelists, and other ministers of Christ. When the sick come to "the elders of the church" for their prayers, it is *"the prayer of faith"* that proves effectual. What is "the prayer of faith"? I think that is explained in verse 16: *"The effectual fervent prayer of a righteous man availeth much,"* or as J. B. Phillips renders it, "Tremendous power is made available through a good man's earnest prayer." If we need further help in understanding what is meant, James gives an Old Testament example, one of the greatest examples of effective praying in the whole of the Bible: "Elias was a man subject to like passions as we are (meaning that he was not a superman, but an ordinary mortal such as we are), and he prayed earnestly that it might not rain and it rained not on the earth by the space of three years and six months. And he prayed again and the heaven gave rain and the earth brought forth her fruit." If we look back at the Old Testament record, we do not find a description of Elijah's praying before the drought, but we do find a most remarkable picture of his praying before the rain came—1 Kings 18:42: "So Ahab went up to eat and to drink. And Elijah went up to the top of Carmel; and he cast himself down upon the earth, and put his face between his knees." What an intriguing picture of earnest prayer! But listen! "And he said to his servant, Go up now, look toward the sea. And he went up, and looked, and said, There is nothing. And he said, Go again seven times. Then he said, Behold, there ariseth a little cloud out of the sea, like a man's hand. And he said, Go up, say unto Ahab, Prepare the chariot, and get thee down, that the rain stop thee not. And it came to pass in the meanwhile, that the heaven was black with clouds and wind, and there was a great rain."

Surely this is what James means by "the prayer of faith." This is "the effectual fervent prayer that availeth much." And this is the kind of praying that is required if the ministers of Jesus Christ are to appropriate and communicate divine healing for others.

Oswald Smith, in his book *The Revival We Need*, says, "Conversion is the operation of the Holy Spirit, and prayer is the power that secures that operation" (p. 24). We may adapt that statement to divine healing. "Healing is the operation of the Holy Spirit, and prayer is the power that secures that operation."

But herein is exposed our weakness and our shame. *There are so few Elijahs*. Too many of us have gone up with Ahab *"to eat and drink."* Too few have gone up with Elijah to "cast ourselves down upon the earth and put our face between our knees" in burdened, agonized, earnest, importunate, believing prayer. Perhaps this is the paramount reason why there are so few releasings of the power of God in the healing of the sick.

In Order To Help People Find Healing We Must Try To Encourage Their Faith and To Increase Our Own Faith

It was Christ Himself who stressed the absolute importance of faith as a condition of appropriating healing for oneself, or of communicating it to others. Let us recall some examples of His teaching: "Verily I say unto you, If ye have faith, and doubt not, ye shall not only do this which is done to the fig tree, but also if ye shall say unto this mountain, Be thou removed, and be thou cast into the sea; it shall be done. And all things, whatsoever ye shall ask in prayer, believing, ye shall receive" (Matthew 21:21-22). To the father of a demon possessed boy who said to Jesus, "If thou canst do anything, have compassion on us and help us," Christ replied, "If thou canst believe! All things are possible to him that believeth" (Mark 9:23, R.V.), and when the disciples asked Him why they were helpless before this case of need, He replied "Because of your little faith: for verily I say unto you, If ye have faith as a grain of mustard seed, ye shall say unto this mountain, Remove hence to yonder place; and it shall remove; and nothing shall be impossible unto you" (Matt. 17:20, R.V.). Matthew describes the healing of two blind men who followed Jesus, crying "Thou Son of David have mercy on us." Before touching their eyes He asked them, "Believe ye, that I am able to do this?" When they said "Yea, Lord," He "touched them saying, According to your faith be it unto you" (Matthew 9:27, 28). Mark describes the healing of Jairus' daughter. As they were going to the house news came that the girl was already dead. "As soon as Jesus heard the word that was spoken, he said unto the ruler of the synagogue, Be not afraid, only believe" (Mark 5:36).

Frequently Jesus commended people for having faith, or remarked upon their faith. Of the Roman Centurion who asked for what is sometimes called "absent healing," He said, "Verily I say unto you, I have not found so great faith, no, not in Israel" (Matthew 8:10). When the woman with an issue of blood touched the hem of His garment, He said to her, "Thy faith hath made thee whole" (Matthew 9:22). It is very evident that faith, equally with prayer, is a fundamental condition of divine healing.

But what is faith? When we study the word in the New Testament we discover that it has a variety of meanings. Sometimes it means little more than hope. Sometimes it means trust. At other times it is a strong conviction or assurance. Sometimes it means fidelity or faithfulness. Once, at least, it is said to be one of the gifts of the Spirit, while nearly twenty times it means the body of Christian truth, "the faith once delivered to the saints."

Now, the faith that appropriates divine healing, it seems to me, is *faith in the sense of trust*. This is the self-same quality of faith that we term "saving faith." We are "saved by grace through faith" (Ephesians 2:8). "Believe on the Lord Jesus Christ and thou shalt be saved" (Acts 16:31). This is not mere intellectual assent to a doctrine. *It is trust in a Person and in a message*. It is a calm and steadfast reliance upon Christ, upon the power of His precious blood for our eternal salvation.

The faith that receives divine healing is a similar kind of faith to the faith that receives salvation. It is well illustrated in the faith of the Virgin Mary, the mother of the Lord. When Elisabeth, the mother of John the Baptist, met her, she exclaimed, *"Blessed is she that believed!"* When we examine the nature of Mary's faith we find that she simply trusted herself into the hands of God. She did not put out some energy or force of faith. She did not think positively. The miracle of the Incarnation and the Virgin Birth was not her idea or her achievement. The entire performance was a divine miracle. Mary simply placed herself in God's hands, and trusted herself absolutely to His working: "Behold the handmaid of the Lord; be it unto me according to thy word" (Luke 1:38).

The faith which receives divine healing is a similar kind and quality of faith. It is trustful reliance upon God's promise; it is putting the same confidence in God's revelation in the written Word as people did in His revelation in the living Word during the Gospel period.

There is a striking example of what this implies in the story of the ten lepers in Luke 17:11-19. These ten segregated sufferers, having heard that the Great Physician was coming their way, were filled with hope that Christ would do something to help them in their misery. Hence their prayer, "Master, have mercy on us!" In this case it is interesting that Jesus did nothing of a specific nature to bring them healing. He did not touch them; there was no so-called "point of contact." He did not even speak a word of healing. "And when he saw them, he said unto them, Go shew yourselves unto the priests." According to Old Testament law the priests were also medical officers who examined skin rashes for signs of leprosy, and pronounced them either *"clean"* or *"unclean,"* as the case might be. These men had already been pronounced "unclean" and segregated from society. But they had the right of further examination from time to time. If there were signs of improvement in his condition a leper could "shew himself to the priests." That is what Jesus told the ten lepers to do. The implication was that there was a change in their condition. When Jesus said, "Go shew yourselves unto the priests," He put the thought in their minds that they were healed, or at least, improved, though He had not yet actually healed them.

And the lepers acted on what He said. They started out for the priests, no doubt going in different directions toward different priests in their own home towns. There was no evidence as yet that they were healed. Yet they went! They simply trusted Christ and obeyed the word. *"And as they went they were cleansed."* As Dr. Campbell Morgan says, *"their going was the going of faith."* There were no signs or feelings of healing in their bodies. Indeed, there were all the marks of their dreadful malady. They simply obeyed the word of the Lord. They accepted unquestioningly the implications of that word. "They went at Christ's bidding, even before they had actually experienced the healing" (Edersheim).

This was not positive thinking. It was not autosuggestion. They were not trying to force themselves to believe that healing had taken, or was taking place. It was simple, childlike, unquestioning action in response to Christ's word. And He honoured their faith by a miracle of healing.

It is this kind and quality of faith which is required for the appropriation of all and every one of the promises of God, not least, the promise of divine healing. It is this kind of unquestioning, undoubting attitude, this total commitment to the Lord, which opens the heart and mind and body to the divine blessing. This is what Jesus meant when He said, "I tell you, then, whatever you ask for in prayer, believe that you have received it, and it will be yours" (Mark 21:24, N.E.B.). It is what John meant when he wrote, "This is the confidence that we have in him, that if we ask anything according to his will, he heareth us: and if we know that he hear us, whatsoever we ask, we know that we have the petitions that we desired of him" (1 John 5:14-15).

Unfortunately, we do not seem always to be able to pray with such confidence and faith, either for divine healing or anything else. Probably one reason for this is that the promises of the written Word of God do not always lay hold of us with the same commanding force as did the utterances of Christ when He stood before men and women in the days of His flesh. I believe this is a matter which we should give attention to. When Christ was among men He spoke with commanding authority. The power of His divine personality riveted His sayings upon the minds of His hearers. It was therefore, comparatively easy to trust Him for a miracle.

Today we do not have His presence in the same way. We have His Word, His message, His Gospel. We are assured that He is both able and willing to do for us what He did then. But somehow faith is not as easy. This may be one reason why miracles of healing are not as frequent. However, there is one compensating factor. *We live in the Pentecostal era.* God has given the Holy Spirit to represent Him in the world and especially in the Church. Part of the great ministry of the Holy Spirit is *to bear witness.* John says, "And it is the Spirit that beareth witness, because the Spirit is truth" (1 John 5:6). This seems to mean that the Holy Spirit takes hold of the historic revelation of Jesus Christ which is embodied in the written Word, vitalizing it, empowering it, quickening it, anointing it, until it glows, as it were, in letters of fire, and grips the mind and heart of the hearer and reader in a similar way that the personal presence of Jesus did in the days of His flesh. *It is when this happens that living faith is created in the heart.* This is so of all the saving truths of the Bible. They are plainly written in the inspired and infallible record—the truth of justification, the truth of regeneration, the truth of sanctification, the truth of the Pentecostal baptism, the truth of divine healing. Yet we can read them and know them and assent to them intellectually, but never experience them in a living way, never appropriate them to ourselves. Though preachers exhort us to believe, there seems no vitality to our faith. Then, it may be, one day the Holy Spirit bears witness to them in our hearts. He makes the truth come alive, and come alive, in particular, *for us*. We feel that the truth is for us, that the promise is ours, and we are able to lay hold of it and appropriate it, and claim it, and experience it.

This, I believe, is how sinners get truly saved. This is how Christians get truly sanctified and filled with the Holy Spirit. This is how the sick are healed. There is a moment when real, living, experiential faith is begotten, and the mighty work is done.

It is the responsibility of Christian pastors and evangelists to try to encourage sick and suffering people to trust in Christ for healing. But we cannot and must not try to manipulate them or force them in any way, or foist upon them any psychological substitute for a living faith in Christ. We must understand that unless the Holy Spirit bears witness to the truths we preach, the hearers can never experience real faith. Therefore, we must not only be preachers of truth but men of prayer. As we have previously said, only prayer can secure the operation of the Holy Spirit, and until there is an operation of the Holy Spirit, men and women cannot truly believe with appropriating faith.

I should like to make a further observation concerning faith for divine healing. I have said that we should try to encourage people to believe for healing. *I believe that we should try also to increase our own faith.* If we are to communicate divine healing to others, that is if we are to be used of the Lord to convey His healing blessing to men and women in need, we ourselves must be men and women of faith.

To illustrate this, let us consider Peter in the case of the healing of the lame man at the Beautiful Gate (Acts 3:1-16). When Peter was explaining what had happened, he declared, "His name through faith in his name hath made this man strong, whom ye see and know: Yea, the faith which is by him hath given him this perfect soundness in the presence of you all" (v. 16). Now it is evident that the faith which brought about the healing miracle was the faith of Peter and John. Not the faith of the lame man! Of course, he too had faith. As Peter spoke to him the Holy Spirit evidently gripped his mind with the promise of healing, and appropriating faith was born. But the real instruments were Peter and John. It was they who had faith in the name of Jesus Christ for the man's healing.

The fact is underscored in the Gospels. When the Apostles asked Jesus why they were so impotent before the demon-possessed boy, He replied, "because of your unbelief [or little faith, R.V.]" (Matthew 17:20). If we are to become instruments in the hands of God to help others believe for divine healing, we must become "strong in faith." This means that we must be Spirit-baptized, Spirit-filled, Spirit-anointed leaders. For just as it takes a work of the Holy Spirit to enable people to have an appropriating faith, so it takes a work of the Spirit to enable people to have a communicating faith.

Three vital qualities are required to make one an effective instrument for the continuance of Christ's healing ministry: 1. We must be men and women of compassion, as Christ Himself was. 2. We must be men and women of prayer, as He was. 3. We must be men and women of faith, as He was. And to make this possible, we must be men and women who live constantly under the anointing of the Holy Spirit, as Christ Himself did.

I Should Like To Comment Briefly Upon the Problem of the Unhealed

I shall not spend much time upon this tremendous problem, as it is the main burden of my book *Sickness, Health and God*. But we cannot and must not ignore it. The problem exists—agonizing and gigantic. However, we seek to explain it, whatever the reason, there are people, many people, often the best and finest of people, who cannot find healing of body. Whether the explanation lies in lack of prayer, failure in faith, wrong motive, or in the sovereignty of God, the plain fact is, there are many people among us who do not get healed. Sometimes people are healed only after many years of suffering and seeking. Others are released only by death.

One of the all-important things in life is how we face up to such a strange and perplexing situation. There have been many people who have been utterly crushed because of it; they have become bitter, resentful, hard. Unanswered prayer and unrequited faith have poisoned the well-springs of their soul, and they have lost their faith.

Others have acted in quite the opposite manner. They have been drawn nearer to God. Their faith has grown stronger and deeper. Their character has become sweeter, cleaner, mellower, finer. This phenomenon has been beautifully expressed in a well-known stanza—

> "Ships sail east and ships sail west
> By the self-same wind that blows.
> It's the set of the sail and not the gale
> That determines the way she goes."

Of course, this is not always as easy as it sounds in poetry. Sometimes, like the ship in which Paul sailed to Rome, we are caught up in a Euroclydon. We are "exceedingly tossed with a tempest," we are "driven" helplessly before the wind, and "neither sun nor stars appear for many days." Yet, in general, it is true that it is "the set of the sail and not the gale" which determines our direction and our destiny.

As I understand it, the one great thing that determines the direction and destiny of our lives, whatever the weather and the circumstances and the ship wrecks, *is a right attitude towards God.* The man who can learn to trust God, and love God, and utterly submit Himself to God in all things, even though his cherished hopes and aspirations are unrealized, even though his prayers seem unanswered, and his faith cannot grasp its objectives—this man will be undefeatable and indestructible.

The Bible abounds in such stories of triumph. It has been my privilege to be acquainted with many of "like precious faith" throughout my life, some of them ministers and leaders, others the rank and file of our church members.

I am sure that the bed-rock of a truly religious and spiritual character is such a loving, steadfast trust in God. The Psalmist declared, "They that trust in the Lord shall be as mount Zion, which cannot be removed, but abideth for ever" (Psa. 125:1). That kind of faith makes that kind of character. And I believe that it is the duty of Christian ministers to help nurture and develop such faith and character in God's people. While we seek, rightly, to encourage them to appropriate the promise of healing, we must accept the fact that many, in spite of much prayer, do not receive healing. We cannot turn aside from such people. We cannot leave them troubled and baffled and soured. We have a responsibility to them. We have to get them on solid rock. We are not merely to encourage appropriating faith. We are to nourish and cherish and foster in them that rock-like faith in God Himself which survives all storms and tempests, all floods and fires, all frustrations and failures, and all apparently unanswered prayers.

Men and women of such faith and such character know in themselves that there is no such thing as unanswered prayer. God never gives a merely negative answer, never a blunt refusal. If He seems to say "No," it is because He purposes to give "some better thing." As Samuel Chadwick says in a profound chapter on "The Problem of Unanswered Prayer," in his wonderful book *The Path of Prayer,* "No inspired prayer of faith is ever refused. *No,* is never God's last word. If prayer seems unanswered, it is because it is lost in the glory of the answer when it comes" (p. 191).

This truly Christian attitude is powerfully expressed by Elizabeth Barrett Browning in her poem on "Unanswered Prayer":

"Unanswered yet! Faith cannot be unanswered;
Her feet are firmly planted on the Rock;

Amid the wildest storms she stands undaunted,
Nor quails before the loudest thunder shock.
She knows Omnipotence has heard her prayer,
And cries, 'It shall be done sometime—somewhere!' "

BIBLIOGRAPHY

WORKS ON DIVINE HEALING . . .

Bingham, Roland J., *The Bible and the Body*, Marshall, Morgan & Scott, Third edition, 1921.
Cobb, Howard, *Christ Healing*, Marshall, Morgan & Scott, 7th impression, 1954.
Edmonds, V. C. & Scorer, C. G., *Some Thoughts on Faith Healing*, The Tyndale Press, Second Edition, 1968.
Frost Evelyn, *Christian Healing*, A. R. Mowbray & Co. Third impression, 1954.
Frost Henry, *Miraculous Healing*, Marshall, Morgan & Scott, 1951.
Jeffreys, George, *Healing Rays*, Elim Publishing Co., 1932.
Kirby, Gilbert W., *The Question of Divine Healing*, A Symposium, Victory Press, 1967.
Murray, Andrew, *Divine Healing*, Victory Press, fifth edition, 1952.
Osborn, T. L., *Healing the Sick*, T. L. Osborn Evangelistic Assoc., 1955.
Parker, Percy G., *Divine Healing*, Victory Press, 1931.
Rose, Louis, *Faith Healing*, Victor Gollanez, 1968.
Short, J. Rendle, *The Bible and Modern Medicine*, Paternoster Perss, 1964.
Simpson, A. B., *The Gospel of Healing*, Marshall, Morgan & Scott, 1915 edition.
Warfield, B. B., *Miracles, Yesterday and Today, Real and Counterfeit*, Eerdmans, 1965.
Weatherhead, Leslie D., *Psychology, Religion and Healing*, Hodder & Stoughton, Second impression, 1952.
Webster, Douglas, *The Healing Christ*, The Highway Press, 1963.
Williams, Brian, *The Significance of the Divine Healing Ministry*, Brian Williams Evangelistic Assoc., 1963.
Woodford, L. F. W., *Divine Healing and the Atonement: A Restatement*, Victoria Institute, 1956.

GENERAL WORKS . . .

Alford, Henry, *Commentary on the Greek New Testament*, 4 Vols., Rivingtons, Third edition, 1856.
Barclay, William, *Daily Study Bible*: John, 2 Vols., Saint Andrew Press, Eighth impression, 1965.
Barclay, William, *New Testament Words*, S.C.M. Press, 1964.
Baron David, *The Servant of Jehovah*, Marshall, Morgan & Scott, Third edition, 1954.
Browning, Elizabeth Barrett, Poem: *Unanswered Prayer*.
Bruce, A. B., *The Miraculous Element in the Gospels*, Hodder & Stroughton, 1902.

BIBLIOGRAPHY

Castiglioni, Arture, *Adventures of the Mind,* Sampson Low, Marston & Co., N. D.

Cranfield, C. B., *St. Mark in the Cambridge Greek Testament,* Cambridge University Press, 1959, 1966.

Cremer, H., *Biblio-Theological Lexicon,* T. & T Clark, 1872.

Chadwick, Samuel, *The Path of Prayer,* Hodder & Stroughton, Thirteenth impression, 1948.

Denney, James, *The Death of Christ,* Hodder & Stroughton, 1909.

Edersheim, *The Life and Times of the Messiah,* Longmans Green & Co., Tenth impression, 1900.

Ellicott, C. J., *A Bible Commentary for English Readers: St. Mark,* Cassel & Co., N. D.

Harris, Sarah, *The Incredible Father Divine,* W. H. Allen, 1954.

James, William, *Varieties of Religious Experience,* Longmans, Green & Co., Thirty-ninth impression, 1941.

Lightfoot, J. B., *Commentary on Philippians,* MacMillan & Co., 1891.

Lloyd-Jones, Martin, *Conversions: Psychological and Spiritual,* Inter-Varsity Fellowship, 1965.

MacGregor, G. H. C., *Moffat Commentary: St. John,* Hodder & Stroughton, N. D.

Morgan, G. Campbell, *The Four Gospels: Luke,* Oliphants, 1956, 1962.

Orr, James, *The Christian View of God and the World,* Andrew Elliot, 1893.

Packard, Vance, *The Hidden Persuaders,* Pelican Books, 1960.

Pentecostal Holiness Church, *Manual of Discipline,* Advocate Press, 1965.

Sargent, William, *Battle for the Mind: A Physiology of Conversion and Brainwashing,* William Heinemann, 1957.

Saunders, J. Oswald, *Heresies and Cults,* Marshall, Morgan & Scott, Revised edition, 1962.

Scofield, A. T., *Nervousness.*

Selbie, W. B., *The Psychology of Religion,* Oxford University Press, 1924.

Smeaton, George, *The Doctrine of the Holy Spirit,* Banner of Truth Trust, 1958.

Smith, G. Abbott, *Manual Greek Lexicon of the New Testament,* T. & T. Clark, Third edition, 1937.

Smith, J. Oswald, *The Revival We Need,* Marshall, Morgan & Scott, Third edition, 1946.

Temple, William, *Readings in St. John's Gospel,* 2 Vols., MacMillan & Co., 1938.

Thomas, W. H. Griffith, *The Principles of Theology,* Church Book Room Press, Fifth revised edition, 1956.

Vine, W. E., *Expository Dictionary of New Testament Words,* Oliphants, Fourteenth impression, 1964.